Problem Books in Mathematics

Edited by P. Winkler

For other titles published in this series, go to
www.springer.com/series/714

Alexander Komech • Andrew Komech

Principles of Partial Differential Equations

Alexander Komech
Faculty of Mathematics
Vienna University
1090 Vienna
Austria
alexander.komech@univie.ac.at

Andrew Komech
Department of Mathematics
Texas A&M University
College Station, TX 77843
USA
comech@math.tamu.edu

Series Editor:
Peter Winkler
Department of Mathematics
Dartmouth College
Hanover, NH 03755
USA
peter.winkler@dartmouth.edu

ISBN 978-1-4614-2462-8   e-ISBN 978-1-4419-1096-7
DOI 10.2007/978-1-4419-1096-7
Springer New York Dordrecht Heidelberg London

Mathematics Subject Classification (2000): 32-XX, 35-00, 35-01

Cover art: Olga Rozmakhova, nominatim.olga@gmail.com

Printed on acid-free paper

Springer is part of Springer Science+Business Media (www.springer.com)

Preface

This book is intended to give the reader an opportunity to master solving PDE problems. Our main goal was to have a concise text that would cover the classical tools of PDE theory that are used in today's science and engineering, such as characteristics, the wave propagation, the Fourier method, distributions, Sobolev spaces, fundamental solutions, and Green's functions. While introductory Fourier method – based PDE books do not give an adequate description of these areas, the more advanced PDE books are quite theoretical and require a high level of mathematical background from a reader. This book was written specifically to fill this gap, satisfying the demand of the wide range of end users who need the knowledge of how to solve the PDE problems and at the same time are not going to specialize in this area of mathematics. Arguably, this is the shortest PDE course, which stretches far beyond common, Fourier method – based PDE texts. For example, [Hab03], which is a common thorough textbook on partial differential equations, teaches a similar set of tools while being about five times longer.

The book is problem-oriented. The theoretical part is rigorous yet short. Sometimes we refer the reader to textbooks that give wider coverage of the theory. Yet, important theoretical details are presented with care, while the hints give the reader an opportunity to restore the arguments to the full rigor. Many examples from physics are intended to keep the book intuitive for the reader and to illustrate the applied nature of the subject.

The book will be useful for any higher-level undergraduate course and for self-study for both graduate and higher-level undergraduate students, and for any specialty in sciences. Its Russian version has been a standard problem-solving manual at Moscow State University since 1988, and is also used by students of St. Petersburg University and Novosibirsk Universities. Its Spanish version is used at Morelia University in Mexico, while the English draft has already been used in Vienna University and at Texas A&M University.

For further reading we recommend [Str92], [Eva98], and [EKS99].

München,   *Alexander Komech*
August 2007   *Andrew Komech*

Acknowledgements

The first author is indebted to Margarita Korotkina for the fortunate suggestion to write this book, to A.F. Filippov, A.S. Kalashnikov, M.A. Shubin, T.D. Ventzel, and M.I. Vishik for checking the first version of the manuscript and for the advice. Both authors are grateful to H. Spohn (Technische Universität, München) and to E. Zeidler (Max-Planck Institute for Mathematics, Leipzig) for their hospitality and support during the work on the book.

Both authors were supported by Institute for Information Transmission Problems (Russian Academy of Sciences). The first author was supported by the Department of Mechanics and Mathematics of Moscow State University, by the Alexander von Humboldt Research Award, FWF Grant P19138-N13, and the Grants of RFBR. The second author was supported by Texas A&M University and by the National Science Foundation under Grants DMS-0621257 and DMS-0600863.

Acknowledgements

Contents

Chapter 1
Hyperbolic equations. Method of characteristics

1 Derivation of the d'Alembert equation

The d'Alembert equation, also called the one-dimensional wave equation,

$$\frac{\partial^2 u}{\partial t^2}(x,t) = a^2 \frac{\partial^2 u}{\partial x^2} + f(x,t), \qquad x \in [0,l], \quad t > 0, \tag{1.1}$$

where $a > 0$ is a constant, describes small transversal oscillations of a stretched string or longitudinal oscillations of an elastic rod.

Let us give a brief derivation of this equation. For a more rigorous argument, see [Vla84, SD64, TS90].

Transversal oscillations of a string

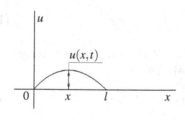

Fig. 1.1

We assume that a string of length l is stretched with the force T. We choose the direction of the axis Ox along the string in its equilibrium configuration. Let $x = 0$ correspond to the left end of the string. Then the right end of the string is given by $x = l$. See Fig. 1.1. We choose the axis Ou normal to Ox, and only consider the transversal oscillations of the string, such that each point x moves only in the direction of the axis Ou. We denote by $u(x,t)$ the displacement of the point x of the string at a moment t.

Alexander Komech and Andrew Komech, *Principles of Partial Differential Equations*,
Problem Books in Mathematics, DOI 10.2007/978-1-4419-1096-7_1,
© Springer Science + Business Media, LLC 2009

We assume that the angles between the string and the axis Ox are small (see Fig. 1.2):

$$|\alpha|, \; |\beta| \ll 1.$$

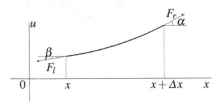

Fig. 1.2

Let us prove that $u(x,t)$ satisfies equation (1.1). To do so, we write Newton's Second Law for a piece of the string from x to $x + \Delta x$, and take its projection onto the axis $0u$:

$$a_u m = F_u. \tag{1.2}$$

Here $a_u \approx \frac{\partial^2 u}{\partial t^2}(x,t)$; $m = \mu \cdot \Delta x$, where μ is the linear density of the string, that is, the mass of its unit length (we assume that the string is uniform), and

$$F_u \approx (F_l)_u + (F_r)_u + \tilde{f}(x,t)\Delta x. \tag{1.3}$$

By F_l and F_r we denoted the force which acts on the region $[x, x + \Delta x]$ from the left and the right part of the string, and $(F_l)_u$, $(F_r)_u$ stand for their projections onto the axis $0u$. $\tilde{f}(x,t)$ is the density of the transversal external forces. For example, in the gravitational field of the Earth, if the string is horizontal and the axis $0u$ is directed upward, then $\tilde{f}(x,t) = -g\mu$, where $g \approx 9.8 \, m/s^2$.

Substituting a_u, m and F_u into (1.2), we obtain

$$\frac{\partial^2 u}{\partial t^2} \mu \Delta x \approx (F_l)_u + (F_r)_u + \tilde{f}(x,t)\Delta x. \tag{1.4}$$

Further, for an elastic string the force of tension T at each point is tangent to the string and has the same magnitude (see [Vla84]). Then

$$(F_l)_u = -T \sin \beta, \qquad (F_r)_u = T \sin \alpha \tag{1.5}$$

and (1.4) takes the form

$$\frac{\partial^2 u}{\partial t^2} \mu \Delta x \approx -T \sin \beta + T \sin \alpha + \tilde{f}(x,t)\Delta x. \tag{1.6}$$

Since we consider the "small" oscillations, such that $|\alpha|$ and $|\beta| \ll 1$, with the precision up to higher powers of α and β,

$$\sin\beta \approx \tan\beta = \frac{\partial u}{\partial x}(x,t), \qquad \sin\alpha \approx \tan\alpha = \frac{\partial u}{\partial x}(x+\Delta x,t). \qquad (1.7)$$

Substituting these expressions into (1.6), we have with the same precision

$$\frac{\partial^2 u}{\partial t^2}\mu\Delta x \approx T\left(\frac{\partial u}{\partial x}(x+\Delta x,t) - \frac{\partial u}{\partial x}(x,t)\right) + \tilde{f}(x,t)\Delta x.$$

Dividing this expression by Δx and sending $\Delta x \to 0$ we obtain equation (1.1), where

$$a = \sqrt{\frac{T}{\mu}}, \qquad f(x,t) = \frac{\tilde{f}(x,t)}{\mu}.$$

Remark 1.1. From our assumption about the tension we deduce that the projections of the forces F_l and F_r onto the axis $0x$ are equal to $-T\cos\beta$ and $T\cos\alpha$, respectively. Therefore their sum $(T\cos\alpha - T\cos\beta)$ is the quantity of magnitude $O(\alpha^2 + \beta^2)$. The projection of the resulting force which acts on the piece of the string from x to $x+\Delta x$ is of the magnitude which is small in the approximation we use. Thus, under this assumption about the tension, the small oscillations of the string are transversal in the precision we use.

Remark 1.2. From (1.5) and (1.7) it follows that

$$T\frac{\partial u}{\partial x}(x,t) \qquad (1.8)$$

is the vertical part of the tension of the string at a point x at a moment t.

Let us consider *the boundary conditions* for the string.

A. If the left end of the string, $x = 0$, is fixed, then its displacement is equal to zero:

$$u(0,t) = 0, \qquad t > 0. \qquad (1.9)$$

B. Assume that the left end of the string is attached to a tiny ring of negligible mass, which can move without friction along a vertical rod (such an end of the string is called a free end). Then the vertical component of the force with which the rod acts on the left end of the string is equal to zero. Therefore, according to Newton's Third Law, the vertical component (1.8) of the force of tension of the string at $x = 0$ is also equal to zero:

$$\frac{\partial u}{\partial x}(0,t) = 0, \qquad t > 0. \qquad (1.10)$$

C. In a more general case, when we attach a mass m to the left end of the string, there is the boundary condition

$$m\frac{\partial^2 u}{\partial t^2}(0,t) = T\frac{\partial u}{\partial x}(0,t), \qquad t > 0. \qquad (1.11)$$

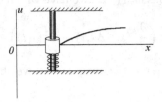

Fig. 1.3

If, besides, the mass m is attached to the spring (as on Fig. 1.3) with the spring constant k, then we need to add the elasticity force $-ku(0,t)$ to the right-hand side of (1.11). If the mass m experiences an additional friction force proportional to the velocity (viscous friction), then one needs to add a friction force $-\eta \frac{\partial u}{\partial t}(0,t)$ to the right-hand side of (1.11). In this way one obtains a physically reasonable linear boundary condition of the form

$$m\frac{\partial^2 u}{\partial t^2}(0,t) = T\frac{\partial u}{\partial x}(0,t) - ku(0,t) - \eta\frac{\partial u}{\partial t}(0,t) + f(t). \qquad (1.12)$$

Here $f(t)$ is an additional external force parallel to the axis Ou which is applied to the left end of the string.

Longitudinal oscillations of an elastic rod

Assume we have a uniform unstretched rod of length l. Choose the axis Ox along the rod, so that its left end is located at the point $x = 0$. Then $x = l$ is its right end. We will consider only longitudinal oscillations of the rod. Denote by $u(x,t)$ the displacement of the point x of the rod along the axis Ox, at the moment t. See Fig. 1.4.

Fig. 1.4

Let us prove that $u(x,t)$ satisfies equation (1.1). For this, we write down the projection onto the axis Ox of Newton's Second Law for the piece of the rod from x to $x + \Delta x$:

$$a_x m = F_x, \qquad a_x \approx \frac{\partial^2 u}{\partial t^2}(x,t), \qquad m = \mu \Delta x. \qquad (1.13)$$

The force F_x has the form

$$F_x = F_l + F_r + \tilde{f}(x,t)\Delta x,$$

where F_l (respectively, F_r) is the force along the axis Ox, acting at the piece $[x, x + \Delta x]$ from the left (respectively, right) piece of the rod, and $\tilde{f}(x,t)$ is the density of the external forces directed along the axis Ox. For example, if the rod is hanging vertically in the field of gravity of the Earth so that the axis Ox is directed downwards, then $\tilde{f}(x,t) = g\mu$.

Substituting F_x into (1.13), we get

$$\frac{\partial^2 u}{\partial t^2}(x,t)\mu\Delta x \approx F_l + F_r + \tilde{f}(x,t)\Delta x. \qquad (1.14)$$

To find F_l and F_r, we use Hook's Law

$$\sigma(x,t) = E\varepsilon(x,t). \qquad (1.15)$$

Here $\sigma(x,t)$ is a tension of the rod at the point x, that is, $\sigma(x,t) = T(x,t)/S$, where $T(x,t)$ is the tension force at the point x and S is the section area; E is Young's module of the material of the rod, and $\varepsilon(x,t)$ is the relative deformation at the point x. For the piece of the rod $[x, x+h]$, its initial length (when no force is applied) is equal to h, while under tension it is $h + u(x+h,t) - u(x,t)$. Therefore the absolute length increase is equal to $u(x+h,t) - u(x,t)$, while the relative length increase is

$$\frac{u(x+h,t) - u(x,t)}{h} \xrightarrow[h \to 0]{} \frac{\partial u}{\partial x}(x,t).$$

Thus,

$$\varepsilon(x,t) = \frac{\partial u}{\partial x}(x,t).$$

From here, by Hook's Law (1.15),

$$T(x,t) = S\sigma(x,t) = SE\varepsilon(x,t) = SE\frac{\partial u}{\partial x}(x,t). \qquad (1.16)$$

Let us point out that Hook's Law (1.15) is a linear approximation for the dependence of $\sigma(x,t)$ of $\varepsilon(x,t)$, and is only applicable for small deformations, that is, small values of $\varepsilon(x,t)$.

Taking into account the direction of the forces F_l and F_r, we obtain:

$$\begin{cases} F_l = -T(x,t) = -SE\frac{\partial u}{\partial x}(x,t), \\ F_r = -T(x+\Delta x,t) = -SE\frac{\partial u}{\partial x}(x+\Delta x,t). \end{cases} \tag{1.17}$$

Indeed, if, for example, $u(x,t)$ is monotonically increasing in x, then the rod is stretched out, hence $F_l \leq 0$, while $F_r \geq 0$. At the same time $\frac{\partial u}{\partial x} \geq 0$. This means that the signs in (1.17) are correct.

Substituting (1.17) into (1.14), we get

$$\frac{\partial^2 u}{\partial t^2}(x,t)\mu\Delta x \approx SE\frac{\partial u}{\partial x}(x+\Delta x,t) - SE\frac{\partial u}{\partial x}(x,t) + \tilde{f}(x,t)\Delta x.$$

From here, dividing by Δx, at the limit $\Delta x \to 0$ we get (1.1) with

$$a = \sqrt{\frac{SE}{\mu}} = \sqrt{\frac{E}{\rho}}, \qquad f(x,t) = \frac{\tilde{f}(x,t)}{\mu},$$

where $\rho = \mu/S$ is the density of the material of the rod.

Let us consider *boundary conditions* for the rod.

a. For the fixed end of the rod at $x = 0$ there is the boundary condition (1.9).

b. For the free end of the rod at $x = 0$, the tension (1.16) is equal to zero. Therefore (1.10) holds.

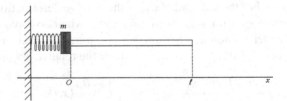

Fig. 1.5

c. In a more general case, assume that there is a mass m at the left end $x = 0$ of the rod, attached to the spring with spring constant $k > 0$, and that the equilibrium position of the spring corresponds to zero displacement of the left end of the rod. See Fig. 1.5. Assume that the mass moves with the viscous friction: $F_{fr} = -\eta v$, where v is the speed of the mass and $\eta > 0$. Then at $x = 0$ there is the boundary condition

$$m\frac{\partial^2 u}{\partial t^2}(0,t) = -ku(0,t) + SE\frac{\partial u}{\partial x}(0,t) - \eta\frac{\partial u}{\partial t}(0,t) + f(t), \tag{1.18}$$

where $f(t)$ is the external force, acting at the left end of the rod along the axis Ox.

2 The d'Alembert method for infinite string

The Cauchy problem for the d'Alembert equation

We consider the d'Alembert equation (1.1) in the real line:

$$\frac{\partial^2 u}{\partial t^2} = a^2 \frac{\partial^2 u}{\partial x^2}, \qquad -\infty < x < \infty, \quad t > 0. \tag{2.1}$$

This corresponds to the physical problem about a string of relatively large size. For simplicity we assume that $f(x,t) \equiv 0$, that is, that there are no external forces.

As we will see below, there are infinitely many solutions of (2.1). To be able to determine the movement of the string it suffices to prescribe initial position and velocity of all points of the string (as usually in mechanics):

$$u(x,0) = \varphi(x), \qquad \frac{\partial u}{\partial t}(x,0) = \psi(x), \quad x \in \mathbb{R}. \tag{2.2}$$

These are *initial conditions* for equation (2.1). Here φ and ψ are prescribed functions, $\varphi(x)$ is initial displacement, and $\psi(x)$ is the initial velocity of a point x of the string.

The problem (2.1)–(2.2) is called the Cauchy problem (or initial value problem) for the d'Alembert equation (2.1). The relations (2.2) are called the boundary conditions, and the functions $\varphi(x)$, $\psi(x)$ are called initial data.

The d'Alembert method

The d'Alembert method is based on the fact that the general solution to (2.1) has the form

$$u(x,t) = f(x - at) + g(x + at), \tag{2.3}$$

where f and g are arbitrary functions of one variable.

Remark 2.1. If f and g belong to $C^2(\mathbb{R})$, then $u(x,t)$ also has two continuous derivatives. It turns out, though, that one can take f and g non-smooth and even discontinuous. Then $u(x,t)$ is also non-smooth or discontinuous, respectively. This is done rigorously in Appendix 27, where we show that such a discontinuous function satisfies equation (2.1) in the sense of distributions.

To prove (2.3), let us rewrite the differential equation (2.1) in the variables

$$r = x - at, \qquad s = x + at. \tag{2.4}$$

Change of variables in a differential equation

Let us express the function $u(x,t)$ in the new coordinates r, s:

$$u(x,t) = v(r,s),$$

where r, s are related to x, t by (2.4). For example,

$$u(x,t) = x \quad \Rightarrow \quad v(r,s) = \frac{1}{2}(r+s).$$

To make a change of variables in the differential equation (2.1) means to find a differential equation on the function $v(r,s)$, which would be equivalent to (2.1). For this we need to express $\frac{\partial^2 u}{\partial t^2}$ and $\frac{\partial^2 u}{\partial x^2}$ via the derivatives of $v(r,s)$ with respect to r, s and to substitute the resulting expressions into (2.1). The necessary expressions are obtained with the aid of the chain rule applied to the identity

$$u(x,t) = v\big(r(x,t), s(x,t)\big). \tag{2.5}$$

Namely, differentiating (2.5) with respect to t and x, we obtain

$$\begin{cases} \dfrac{\partial u}{\partial t} = \dfrac{\partial v}{\partial r}\dfrac{\partial r}{\partial t} + \dfrac{\partial v}{\partial s}\dfrac{\partial s}{\partial t}, \\[2mm] \dfrac{\partial u}{\partial x} = \dfrac{\partial v}{\partial r}\dfrac{\partial r}{\partial x} + \dfrac{\partial v}{\partial s}\dfrac{\partial s}{\partial x}. \end{cases} \tag{2.6}$$

In the same way one can express all other derivatives. Differentiating the first relation (2.6) with respect to t, we obtain:

$$\frac{\partial^2 u}{\partial t^2} = \left(\frac{\partial}{\partial t}\frac{\partial v}{\partial r}\right)\frac{\partial r}{\partial t} + \frac{\partial v}{\partial r}\frac{\partial^2 r}{\partial t^2} + \left(\frac{\partial}{\partial t}\frac{\partial v}{\partial s}\right)\frac{\partial s}{\partial t} + \frac{\partial v}{\partial s}\frac{\partial^2 s}{\partial t^2}. \tag{2.7}$$

We express the operator $\frac{\partial}{\partial t}$ from the same relation (2.6):

$$\frac{\partial}{\partial t} = \frac{\partial r}{\partial t}\frac{\partial}{\partial r} + \frac{\partial s}{\partial t}\frac{\partial}{\partial s}.$$

Substituting this expression in (2.7), we get

$$\begin{aligned} \frac{\partial^2 u}{\partial t^2} &= \left(\frac{\partial r}{\partial t}\frac{\partial^2 v}{\partial r^2} + \frac{\partial s}{\partial t}\frac{\partial^2 v}{\partial s\,\partial r}\right)\frac{\partial r}{\partial t} + \frac{\partial v}{\partial r}\frac{\partial^2 r}{\partial t^2} \\[2mm] &\quad + \left(\frac{\partial r}{\partial t}\frac{\partial^2 v}{\partial r\,\partial s} + \frac{\partial s}{\partial t}\frac{\partial^2 v}{\partial s^2}\right)\frac{\partial s}{\partial t} + \frac{\partial v}{\partial s}\frac{\partial^2 s}{\partial t^2} \\[2mm] &= \left(\frac{\partial r}{\partial t}\right)^2\frac{\partial^2 v}{\partial r^2} + 2\frac{\partial r}{\partial t}\frac{\partial s}{\partial t}\frac{\partial^2 v}{\partial r\,\partial s} + \left(\frac{\partial s}{\partial t}\right)^2\frac{\partial^2 v}{\partial s^2} + \frac{\partial v}{\partial r}\frac{\partial^2 r}{\partial t^2} + \frac{\partial v}{\partial s}\frac{\partial^2 s}{\partial t^2}. \end{aligned} \tag{2.8}$$

Here we used the identity

$$\frac{\partial^2 v}{\partial s \partial r} = \frac{\partial^2 v}{\partial r \partial s}.$$

In the same fashion (substituting in (2.8) t by x) one can obtain the formula

$$\frac{\partial^2 u}{\partial x^2} = \left(\frac{\partial r}{\partial x}\right)^2 \frac{\partial^2 v}{\partial r^2} + 2\frac{\partial r}{\partial x}\frac{\partial s}{\partial x}\frac{\partial^2 v}{\partial r \partial s} + \left(\frac{\partial s}{\partial x}\right)^2 \frac{\partial^2 v}{\partial s^2} + \frac{\partial v}{\partial r}\frac{\partial^2 r}{\partial x^2} + \frac{\partial v}{\partial s}\frac{\partial^2 s}{\partial x^2}.$$

Problem 2.2. Derive the relation

$$\frac{\partial^2 u}{\partial t \partial x} = \frac{\partial r}{\partial t}\frac{\partial r}{\partial x}\frac{\partial^2 v}{\partial r^2} + \left(\frac{\partial r}{\partial t}\frac{\partial s}{\partial x} + \frac{\partial s}{\partial t}\frac{\partial r}{\partial x}\right)\frac{\partial^2 v}{\partial r \partial s}$$
$$+ \frac{\partial s}{\partial t}\frac{\partial s}{\partial x}\frac{\partial^2 v}{\partial s^2} + \frac{\partial^2 r}{\partial t \partial x}\frac{\partial v}{\partial r} + \frac{\partial^2 s}{\partial t \partial x}\frac{\partial v}{\partial s}. \qquad (2.9)$$

Remark 2.3. Usually the formulae (2.6) and (2.8)–(2.9) are written with u instead of v. For example, (2.6) is written as

$$\begin{cases} \dfrac{\partial u}{\partial t} = \dfrac{\partial u}{\partial r}\dfrac{\partial r}{\partial t} + \dfrac{\partial u}{\partial s}\dfrac{\partial s}{\partial t}, \\[3mm] \dfrac{\partial u}{\partial x} = \dfrac{\partial u}{\partial r}\dfrac{\partial r}{\partial x} + \dfrac{\partial u}{\partial s}\dfrac{\partial s}{\partial x}. \end{cases} \qquad (2.10)$$

If so, the symbol $\frac{\partial u}{\partial r}$ (respectively, $\frac{\partial u}{\partial s}$) in the right-hand side is to be understood as the derivative along the line $s = \text{const}$ (or $r = \text{const}$):

$$\frac{\partial u}{\partial r} \equiv \frac{d}{dr} u\Big|_{s=\text{const}}, \qquad (2.11)$$

which is actually $\frac{\partial v}{\partial r}$ (respectively, $\frac{\partial v}{\partial s}$), not as "partial derivative of $u(x,t)$ with respect to r (or s)"; the latter does not make sense until the other variable, s (or r), is chosen. Indeed, from (2.11) one can see that $\frac{\partial u}{\partial r}$ depends not only on the choice of the variable r, but also on the variable s, although this is not reflected in the notation $\frac{\partial u}{\partial r}$. Thus, the usage of the notation u instead of v in the right hand side of (2.6) (as this was done in (2.10)) could lead to a confusion.

Problem 2.4. Find $\frac{\partial u}{\partial r}$ for $u(x,t) = t$, $r = x$, and $s = t + x$.

Solution. $t = s - x = s - r$, hence $\frac{\partial u}{\partial r} = -1$.

Problem 2.5. Find $\frac{\partial u}{\partial r}$ for $u(x,t) = t$, $r = x$, and $s = t - x$.

Solution. $t = s + x = s + r$, hence $\frac{\partial u}{\partial r} = 1$.

Nevertheless, in the applied problems, the formulae like (2.10) are often used to avoid the introduction of new notations. For example, the pressure is usually denoted by p, the current is denoted by j, the density is denoted by ρ, etc. We will also use formulae like (2.10) everywhere below.

Proof of the d'Alembert representation (2.3)

From the generic formulae (2.10) for the change of variables (2.4) we derive

$$\frac{\partial}{\partial t} = -a\frac{\partial}{\partial r} + a\frac{\partial}{\partial s},$$

$$\frac{\partial}{\partial x} = \frac{\partial}{\partial r} + \frac{\partial}{\partial s}. \tag{2.12}$$

From this we obtain:

$$\begin{cases} \dfrac{\partial^2}{\partial t^2} = a^2\dfrac{\partial^2}{\partial r^2} - 2a^2\dfrac{\partial^2}{\partial r\partial s} + a^2\dfrac{\partial^2}{\partial s^2}, \\[2mm] \dfrac{\partial^2}{\partial x^2} = \dfrac{\partial^2}{\partial r^2} + 2\dfrac{\partial^2}{\partial r\partial s} + \dfrac{\partial^2}{\partial s^2}. \end{cases} \tag{2.13}$$

Substituting (2.13) into (2.1), we get

$$\left(a^2\frac{\partial^2}{\partial r^2} - 2a^2\frac{\partial^2}{\partial r\partial s} + a^2\frac{\partial^2}{\partial s^2} \right)u = a^2\left(\frac{\partial^2}{\partial r^2} + 2\frac{\partial^2}{\partial r\partial s} + \frac{\partial^2}{\partial s^2} \right)u.$$

After mutual cancellations we obtain

$$\frac{\partial^2 u}{\partial r\partial s} = 0. \tag{2.14}$$

This is the canonical form of the d'Alembert equation (2.1).

The d'Alembert equation in the canonical form (2.14) is easily solved. Denote

$$\frac{\partial u}{\partial s}(r,s) = v(r,s). \tag{2.15}$$

Then (2.14) can be written as

$$\frac{\partial v}{\partial r} \equiv \frac{d}{dr}v\Big|_{s=\text{const}} = 0.$$

It then follows that $v\big|_{s=\text{const}}$ does not depend on r, that is,

$$v(r,s) \equiv c(s),$$

or, taking into account (2.15),

$$\frac{d}{ds}u\Big|_{r=\text{const}} = c(s).$$

Integrating this ordinary differential equation, we obtain

$$u\big|_{r=\text{const}} = \int c(s)\,ds + c_1(r).$$

Thus,

$$u = f(r) + g(s), \tag{2.16}$$

where f and g are some functions of one variable. On the other hand, a function of the form (2.16) satisfies equation (2.14) for any f and g. At last, changing the variables in (2.16) according to (2.4), we obtain the d'Alembert representation (2.3).

Remark 2.6. The graph of a function $f(x - at)$ in (2.4) is a wave moving along the direction of the axis Ox to the right with the speed a, while $g(x + at)$ represents a wave moving with the same speed to the left. This means that the graph of the function $f(x - at)$ (respectively, $g(x + at)$) for any $t > 0$ as a function of x is obtained from the graph of the function $f(x)$ (respectively, $g(x)$) with the aid of a parallel transform to the right (respectively, to the left) along the axis Ox by the distance at. Therefore, the form of the graph of the function $f(x - at)$ considered as a function of x with fixed t is the same. In physics, such functions are called traveling waves. Thus, the d'Alembert decomposition (2.3) means that any solution of the d'Alembert equation is the sum (physicists also use words superposition and interference) of two traveling waves.

The Cauchy problem. The d'Alembert formula

We apply the d'Alembert method to the problem (2.1)–(2.2). To do so, we substitute equation (2.1) with its equivalent (2.3). Thus, we are left to take into account the initial conditions (2.2). It is from these conditions that we will determine the unknown functions f and g from the given φ and ψ.

Namely, substitute (2.3) into (2.2):

$$\begin{cases} f(x) + g(x) = \varphi(x), \\ f'(x)(-a) + g'(x)a = \psi(x), \quad x \in \mathbb{R}. \end{cases} \tag{2.17}$$

Remark 2.7. In the second equation (2.17) we have used the chain rule:

$$\left(\frac{\partial}{\partial t} f(x - at) \right)\bigg|_{t=0} = \left(f'(x - at) \frac{\partial}{\partial t}(x - at) \right)\bigg|_{t=0} = f'(x)(-a).$$

Here $f'(x)$ is an ordinary derivative (not a partial one). This is an important feature of the d'Alembert method: it allows us to reduce the equations (2.1)–(2.2) with partial derivatives to the equations (2.17) with ordinary derivatives.

Integrating the second equation in (2.17) and dividing it by a, we obtain:

$$-f(x) + g(x) = \frac{1}{a} \int_0^x \psi(y)\,dy + \frac{c}{a}.$$

Taking the sum of this equation with the first equation from (2.17) and dividing by two, we obtain

$$g(x) = \frac{1}{2}\varphi(x) + \frac{1}{2a} \int_0^x \psi(y)\,dy + \frac{c}{2a}. \tag{2.18}$$

Instead, taking the difference of these two equations, we arrive at

$$f(x) = \frac{1}{2}\varphi(x) - \frac{1}{2a} \int_0^x \psi(y)\,dy - \frac{c}{2a} = \frac{1}{2}\varphi(x) + \frac{1}{2a} \int_x^0 \psi(y)\,dy - \frac{c}{2a}. \tag{2.19}$$

Substituting these expressions into the d'Alembert decomposition (2.3), we obtain the d'Alembert formula:

$$u(x,t) = \frac{\varphi(x-at) + \varphi(x+at)}{2} + \frac{1}{2a} \int_{x-at}^{x+at} \psi(y)\,dy. \tag{2.20}$$

Remark 2.8. One can see from (2.18)–(2.19) that the waves $f(x-at)$ and $g(x+at)$ are determined by the initial data φ and ψ not uniquely, but only up to an additive constant. At the same time, the solution $u(x,t)$ to the Cauchy problem is uniquely defined.

3 Analysis of the d'Alembert formula

Propagation of waves

Problem 3.1. Let the string be described by (2.1) with $a = 1$. In (2.2), take the initial data as on Fig. 3.1 (see Remark 2.1). Draw the string at $t = 1, 2, 3, 4, 5$.

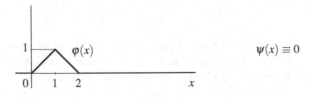

Fig. 3.1

Remark 3.2. We may assume that $\varphi(x)$ is a piecewise-linear function. Then the solution is also going to be a piecewise-linear function, which is a solution of equation (2.1) in the sense of distributions (see Remark 2.1).

Or instead one can think that the graph φ is slightly smoothed out at the corner points, so that $\varphi(x) \in C^2(\mathbb{R})$. Then the solution is also going to be of class C^2, and one should think that all the corners are slightly smoothed out at all the drawings below.

Solution. According to the d'Alembert formula (2.20),

$$u(x,t) = \frac{\varphi(x-t) + \varphi(x+t)}{2}.$$

This means that the graph $\varphi(x)$ should be compressed to the axis Ox by the factor of 2, shifted to the right by t, to the left by t, and the results added up (see Fig. 3.2). Thereafter these humps of hight $\frac{1}{2}$ and width 2 propagate to the left and to the right, each with the speed 1.

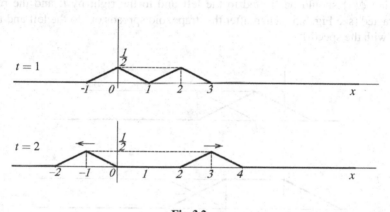

Fig. 3.2

Problem 3.3. In the settings of Problem 3.1, draw the string at $t = \frac{1}{4}$ and $t = \frac{1}{2}$.

Problem 3.4 (Hitting the string with a hammer). Let the string be described by (2.1) with $a = 1$. In (2.2), take the initial data as on Fig. 3.3. Draw the shape of the string at $t = 1, 2, 3, 4, 5$.

Solution. According to the d'Alembert formula (2.20),

$$u(x,t) = \frac{1}{2} \int_{x-t}^{x+t} \psi(y)\,dy = \phi(x+t) - \phi(x-t),$$

where $\phi(x) = \frac{1}{2} \int_0^x \psi(y)\,dy$. See Fig. 3.4. This formula means that the graph of the

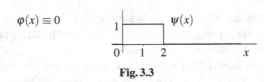

Fig. 3.3

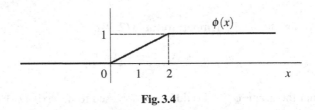

Fig. 3.4

function $\phi(x)$ should be shifted to the left and to the right by t, and the results subtracted (see Fig. 3.5). Thereafter this trapezoid spreads out to the left and to the right with the speed 1.

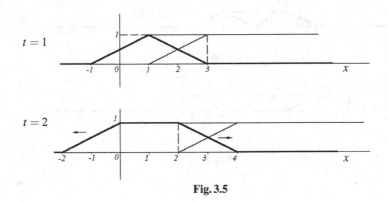

Fig. 3.5

Problem 3.5. In the settings of Problem 3.4, draw the string at $t = \frac{1}{4}$ and $t = \frac{1}{2}$.

Characteristics

When solving two previous problems, we have seen that the lines $x \pm t = \text{const}$ play a special role. For example, the corner points of the graphs of the solutions $u(x,t)$ lie on the lines $x \pm t = 0$ and $x \pm t = 2$.

For equation (1.1) with the coefficient a the similar role is played by the lines $x \pm at =$ const. They are called the *characteristic curves* (or simply *characteristics*) of equation (1.1). Thus, the characteristics of the d'Alembert equation are two families of lines (see Fig. 3.6). We will call the lines $x - at =$ const the characteristics moving to the right (with the speed a). Obviously, they are the level curves of the wave $f(x - at)$. Similarly, the lines $x + at =$ const are called characteristics moving to the left. They are the level curves of the wave $g(x + at)$. The greater the speed a, the smaller is the angle between the characteristics and the axis Ox (if the scale on the axes Ox and Ot is the same, then $\tan \alpha = 1/a$).

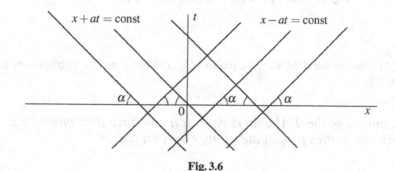

Fig. 3.6

Discontinuities of the solution

Consider a discontinuous function $f(x) = \begin{cases} 0, & x < 2 \\ 1, & x \geq 2 \end{cases}$ as shown on Fig. 3.7.

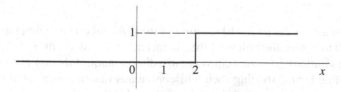

Fig. 3.7

Then the function

$$u(x,t) = f(x - at) \tag{3.1}$$

is discontinuous along the characteristic curve $x - at = 2$. See Fig. 3.8. The func-

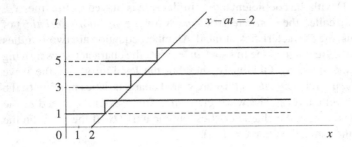

Fig. 3.8 Profiles of the function $u(x,t)$ at $t = 1, 3, 5$.

tion (3.1) satisfies the d'Alembert equation (2.1) in the sense of distributions (see Remark 2.1).

Thus:

> *Solutions of the d'Alembert equation can have discontinuities.*
> *Discontinuities propagate along characteristics.*

Remark 3.6. One can take a smooth function $f_\varepsilon(x)$, which changes from 0 to 1 on a small interval from $x = 2$ to $x = 2 + \varepsilon$, where $\varepsilon > 0$. See Fig. 3.9. Then the function

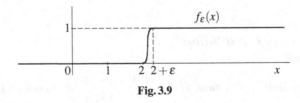

Fig. 3.9

$f_\varepsilon(x - at)$ will be a classical (smooth) solution of the d'Alembert equation, rapidly changing from 0 to 1 near the points of the characteristic curve $x - at = 2$. In the limit $\varepsilon \to 0_+$ the solutions $f_\varepsilon(x - at)$ converge to a discontinuous function $f(x - at)$. This is the meaning behind treating such a discontinuous function as a solution of d'Alembert equation in the sense of distributions (see also Remark 2.1).

Remark 3.7. Discontinuous solutions $u(x,t)$ to the d'Alembert equation for the string and for the rod do not make a physical sense. Still, the d'Alembert equation also describes the gas pressure $p(x,t)$ in a long narrow pipe (such as a flute or an organ; see Fig. 3.10). The function $p(x,t)$ can be discontinuous.

Discontinuous solutions in the dynamics of gas are called the shock waves. When the plane travels with the supersonic speed, there is such a shock wave coming from

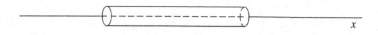

Fig. 3.10

the front edge of the wings, with the pressure being higher behind the front of this wave than ahead of it. We hear a bang when the wave front passes our ears; see Fig. 8.6 in Section 8 below.

Domain of dependence and its graphical representation

Question 3.8. What do we need to know to be able to compute the solution u of the problem (2.1)–(2.2) at the point (x_0, t_0)?

Answer. From the d'Alembert formula (2.20) we see that one needs the initial displacements $\varphi(x)$ at two points: $x = x_0 + at_0$ and $x = x_0 - at_0$, and also the initial velocities $\psi(x)$ on the interval $[x_0 - at_0, x_0 + at_0]$ between these points. Knowing $\varphi(x)$ and $\psi(x)$ beyond the interval $[x_0 - at_0, x_0 + at_0]$ is not needed.

The interval $[x_0 - at_0, x_0 + at_0]$ is called the domain of dependence for the solution to the Cauchy problem (2.1)–(2.2) at the point (x_0, t_0).

Remark 3.9. Now we can explain precisely when we can treat the string as infinite: When the point under consideration x_0 is located at a distance larger than at_0 from the endpoints of the string, where t_0 is the moment of time that we are interested in.

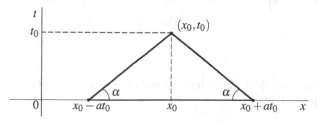

Fig. 3.11

For the graphical representation of the domain of dependence for a solution at (x_0, t_0) we draw two characteristics from this point, as on Fig. 3.11. The intersection of these characteristics with the axis Ox are the points $x_0 - at_0$ and $x_0 + at_0$. Let us check this. The equations of the characteristics are

$$x - at = c_1, \qquad x + at = c_2. \tag{3.2}$$

Since the point (x_0, t_0) lies on these characteristics, $x_0 - at_0 = c_1$ and $x_0 + at_0 = c_2$. To find the intersection of the characteristics with the axis Ox, we need to set $t = 0$ in (3.2), getting

$$x = c_1 = x_0 - at_0 \qquad \text{and} \qquad x = c_2 = x_0 + at_0.$$

The interval of the axis Ox between $x_0 - at_0$ and $x_0 + at_0$ is the domain of dependence for the solution u at the point (x_0, t_0).

Propagation of waves

Problem 3.10. Let $\varphi(x) = \psi(x) = 0$ for $x \notin [2, 5]$. Find the region where the solution $u(x,t)$ to the problem (2.1)–(2.2) is equal to zero for $t > 0$.

Solution. From the points 2 and 5 of the Ox axis we draw the characteristics to the left and to the right, respectively. See Fig. 3.12. In the region below these characteristics the solution is equal to zero. Indeed, for the point (x_0, t_0) below these characteristics the domain of dependence does not intersect the interval $[2, 5]$, hence $\varphi(x) \equiv \psi(x) \equiv 0$ in this domain of dependence. Consequently, $u(x_0, t_0) = 0$.

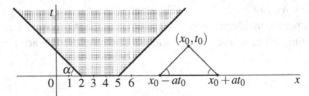

Fig. 3.12

The region above these characteristics is called *domain of influence* for the initial data on the interval $[2, 5]$.

4 Second-order hyperbolic equations in the plane

Factorization of the d'Alembert operator

Let us bring the d'Alembert equation to the canonical form (2.14) using a new method. For this, we rewrite the equation as

$$\Box u \equiv \left(\frac{\partial^2}{\partial t^2} - a^2 \frac{\partial^2}{\partial x^2} \right) u = 0. \tag{4.1}$$

The operator $\Box$ is called the d'Alembert operator, or d'Alembertian. We decompose this operator into factors:

$$\Box u \equiv \left(\frac{\partial}{\partial t} - a \frac{\partial}{\partial x} \right) \left(\frac{\partial}{\partial t} + a \frac{\partial}{\partial x} \right) u = 0. \tag{4.2}$$

The operators

$$L_{(-a,1)} \equiv \frac{\partial}{\partial t} - a \frac{\partial}{\partial x}, \qquad L_{(a,1)} \equiv \frac{\partial}{\partial t} + a \frac{\partial}{\partial x} \tag{4.3}$$

are differentiations along the vectors $(-a, 1)$ and $(a, 1)$, respectively. These vectors are directed along the characteristics

$$x + at = \text{const} \quad \text{and} \quad x - at = \text{const}. \tag{4.4}$$

If we take the characteristic lines as new coordinate axes, setting

$$r = x - at, \qquad s = x + at, \tag{4.5}$$

then due to (2.12) the d'Alembert operator takes the form

$$\Box = \frac{\partial^2}{\partial t^2} - a^2 \frac{\partial^2}{\partial x^2} = L_{(-a,1)} \cdot L_{(a,1)} = -2a \frac{\partial}{\partial r} 2a \frac{\partial}{\partial s} = -4a^2 \frac{\partial^2}{\partial r \partial s}.$$

Conclusion. The characteristics of equation (4.1) are the lines such that the operators of differentiating along them, $L_{(\mp a,1)}$, are the factors of the d'Alembert operator.

Remark 4.1. Since the operators of differentiating along the characteristics are the factors of the d'Alembert operator, this operator sends to zero any function which is constant along characteristics of one of the families (in particular, any such function that is discontinuous; see Remark 2.1). This explains why the solutions to the d'Alembert equation can have discontinuities along characteristics.

Hyperbolic equations with constant coefficients

We consider the equation of the form

$$a\frac{\partial^2 u}{\partial t^2} + 2b\frac{\partial^2 u}{\partial t \partial x} + c\frac{\partial^2 u}{\partial x^2} = 0, \qquad x \in \mathbb{R}, \quad t > 0. \tag{4.6}$$

In this section we assume that the coefficients a, b, and c are constants.

Let us try to apply the method of Section 1 to equation (4.6) instead of (4.1). To obtain factorization like (4.2), we need to decompose into linear factors the "characteristic" quadratic form

$$A(\xi, \tau) \equiv a\tau^2 + 2b\tau\xi + c\xi^2 = \xi^2\left(a\left(\frac{\tau}{\xi}\right)^2 + 2b\frac{\tau}{\xi} + c\right).$$

To achieve this, we solve the characteristic equation

$$a\lambda^2 + 2b\lambda + c = 0. \tag{4.7}$$

Its roots

$$\lambda_{1,2} = \frac{b \pm \sqrt{b^2 - ac}}{a}$$

are real and different if the discriminant is positive:

$$D \equiv b^2 - ac > 0. \tag{4.8}$$

We assume that in this section (4.8) is satisfied. This is the strict hyperbolicity condition for equation (4.6). According to the Viet theorem,

$$a\lambda^2 + 2b\lambda + c = a(\lambda - \lambda_1)(\lambda - \lambda_2).$$

Therefore, the quadratic form turns into

$$A(\xi, \tau) = \xi^2 a\left(\frac{\tau}{\xi} - \lambda_1\right)\left(\frac{\tau}{\xi} - \lambda_1\right) = a(\tau - \lambda_1\xi)(\tau - \lambda_2\xi).$$

Accordingly, the differential equation (4.6) takes the form

$$\left(\frac{\partial}{\partial t} - \lambda_1\frac{\partial}{\partial x}\right)\left(\frac{\partial}{\partial t} - \lambda_2\frac{\partial}{\partial x}\right)u = 0. \tag{4.9}$$

Denote

$$L_{(-\lambda_1, 1)} = \frac{\partial}{\partial t} - \lambda_1\frac{\partial}{\partial x} \quad \text{and} \quad L_{(-\lambda_2, 1)} = \frac{\partial}{\partial t} - \lambda_2\frac{\partial}{\partial x}.$$

Analogously to (4.5), we set

$$r = x + \lambda_1 t, \qquad s = x + \lambda_2 t. \tag{4.10}$$

Then

$$L_{(-\lambda_1,1)}r \equiv 0, \qquad L_{(-\lambda_2,1)}s \equiv 0.$$

Here $L_{(-\lambda_1,1)}$ is the operator of differentiation along the lines $r = $ const, while $L_{(-\lambda_2,1)}$ differentiates along the lines $s = $ const. Hence,

$$L_{(-\lambda_1,1)} = c_1 \frac{\partial}{\partial s}\bigg|_{r=\text{const}}, \qquad L_{(-\lambda_2,1)} = c_2 \frac{\partial}{\partial r}\bigg|_{s=\text{const}}.$$

It follows that (4.9) is equivalent to the equation

$$\frac{\partial}{\partial s}\frac{\partial}{\partial r}u = 0. \tag{4.11}$$

Similarly to (2.16), the general solution to equation (4.6) is given by

$$u = f(r) + g(s) = f(x + \lambda_1 t) + g(x + \lambda_2 t). \tag{4.12}$$

The wave $f(x + \lambda_1 t)$ propagates along the axis x with the speed λ_1, while the wave $g(x + \lambda_2 t)$ propagates with the velocity λ_2 (both waves propagate to the left if $\lambda_1 > 0$, $\lambda_2 > 0$).

In particular, for the d'Alembert equation (4.1), the characteristic equation (4.7) takes the form

$$\lambda^2 - a^2 = 0,$$

so that $\lambda_1 = -a$, $\lambda_2 = a$, and (4.10) turns into (2.4), while (4.12) turns into (2.16).

With the aid of the representation (4.12), all the conclusions of Section 3 about discontinuities of the solution, propagation of waves, and the regions of dependence are easily generalized for equation (4.6) (see Remark 2.1).

The characteristics of equation (4.6) are defined by relations

$$r \equiv x + \lambda_1 t = \text{const}, \qquad s \equiv x + \lambda_2 t = \text{const}. \tag{4.13}$$

Solutions of equation (4.6) may have singularities along these characteristics. This is seen from (4.12) in the case when f or g are not smooth (see also Remark 4.1).

The Cauchy problem (4.6) with initial data (2.2) has a solution

$$u(x,t) = \frac{\lambda_2 \varphi(x + \lambda_1 t) - \lambda_1 \varphi(x + \lambda_2 t)}{\lambda_2 - \lambda_1} + \frac{1}{\lambda_2 - \lambda_1} \int_{x+\lambda_1 t}^{x+\lambda_2 t} \psi(y)\,dy. \tag{4.14}$$

Problem 4.2. Derive the formula (4.14).

Let us point out that for the d'Alembert equation one has $\lambda_1 = -a$, $\lambda_2 = a$, so that (4.14) turns into the d'Alembert formula (2.20).

As seen from (4.14), the domain of dependence for the solution u at a point (x_0, t_0) is the interval $[x_0 + \lambda_1 t_0, x_0 + \lambda_2 t_0])$ of the axis Ox. Its ends are the intersection points of the axis Ox with characteristics (4.13) sent back in time from the point (x_0, t_0); see Fig. 4.1.

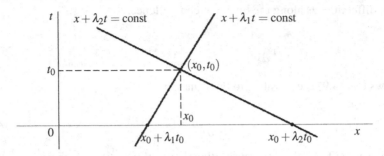

Fig. 4.1

Let us point out that the roots λ_1 and λ_2 could be of the same sign; then the waves $f(x+\lambda_1 t)$ and $g(x+\lambda_2 t)$ run into the same direction.

Example 4.3. For the equation

$$\left(\frac{\partial^2}{\partial t^2} + 5\frac{\partial^2}{\partial t \partial x} + 6\frac{\partial^2}{\partial x^2}\right)u = 0$$

the characteristic equation

$$\lambda^2 + 5\lambda + 6 = 0$$

has the roots $\lambda_1 = -2$, $\lambda_2 = -3$, and the general solution

$$u = f(x-2t) + g(x-3t)$$

consists of two waves propagating to the right.

Let us find the differential equation of the characteristics of equation (4.6).

We note that, according to (4.13), the (co)tangent vectors (dx, dt) to characteristic curves satisfy the equation

$$dx + \lambda_1 \, dt = 0 \quad \text{or} \quad dx + \lambda_2 \, dt = 0.$$

Therefore, either $\frac{dx}{dt} = -\lambda_1$, or $\frac{dx}{dt} = -\lambda_2$; that is, $\lambda \equiv -\frac{dx}{dt}$ satisfies the characteristic equation (4.7):

$$a\left(\frac{dx}{dt}\right)^2 - 2b\frac{dx}{dt} + c = 0.$$

This is the differential equation of the characteristics. It can be written in the following symmetric form:

$$a\,dx^2 - 2b\,dx\,dt + c\,dt^2 = 0. \tag{4.15}$$

Equations with variable coefficients

Now let the coefficients a, b, and c in (4.6) be variable, that is, functions of x and t:

$$a(x,t)\frac{\partial^2 u}{\partial t^2} + 2b(x,t)\frac{\partial^2 u}{\partial t \partial x} + c(x,t)\frac{\partial^2 u}{\partial x^2} = 0; \quad x \in \mathbb{R}, \quad t > 0. \tag{4.16}$$

We will try to generalize the method of Section 2 in order to bring (4.16) to the canonical form (4.11) or at least to a form close to it.

In a small neighborhood of each point (x,t), we substitute equation (4.16) by equation (4.6) with constant coefficients, equal to the values of the coefficients of equation (4.16) at this particular point (x,t). This procedure is called "the freezing of the coefficients."

If we do so, the characteristics of the equation "frozen" at the point (x,t) will have directions which depend on (x,t). The vectors (dx, dt) tangent to these characteristics will satisfy equation (4.15). See Fig. 4.2.

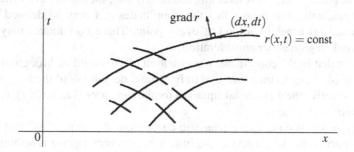

Fig. 4.2

Integral curves of equation (4.15) are called the *characteristic curves* (or simply *characteristics*) of equation (4.16). Thus, due to (4.15), the differential equation of the characteristics of equation (4.16) is given by

$$a(x,t)\,dx^2 - 2b(x,t)\,dx\,dt + c(x,t)\,dt^2 = 0. \tag{4.17}$$

The characteristic equation (4.17) is obtained by a formal substitution

$$\frac{\partial}{\partial t} \mapsto dx, \qquad \frac{\partial}{\partial x} \mapsto -dt. \tag{4.18}$$

Assume that in the region of the (x,t)-plane where we are to solve equation (4.16) the strict hyperbolicity condition (4.8) is satisfied:

$$b^2(x,t) - a(x,t)c(x,t) > 0. \tag{4.19}$$

Then, dividing equation (4.17) by dt^2 and solving the resulting quadratic equation, we obtain two different differential equations:

$$\frac{dx}{dt} = \frac{b \pm \sqrt{b^2 - ac}}{a}. \tag{4.20}$$

If the functions a, b, and c are smooth, equation (4.20) has two corresponding families of the integral curves. We will denote the corresponding families of characteristics by the signs "+" "−," respectively. In the (x,t)-plane, we introduce new coordinates r and s so that $r = \text{const}$ on the characteristics of the family "+" while $s = \text{const}$ on the characteristics of the family "−." This means that the characteristics are the new coordinate curves, and r, s are the first integrals of equations (4.20).

Let us mention that the change of variables $(x,t) \mapsto (r,s)$ is nondegenerate at each point where the condition (4.19) is satisfied. Indeed, one can see from (4.20) that at each point the characteristics have different directions, and since $\text{grad}\, r$ and $\text{grad}\, s$ are orthogonal to the corresponding characteristics, they also have different directions. Incidentally, this means that the coordinates r, s may be defined in a sufficiently small open neighborhood of every point. These coordinates may not exist in the whole region under consideration.

Let us check that in the coordinates r, s equation (4.16) could be brought to the canonical form (4.11) up to the terms that only contain derivatives of the first order. We first need to derive the differential equation for the functions $r(x,t), s(x,t)$, called the *characteristic equation*.

Since $r(x,t) = \text{const}$ on any characteristic curve from the family "+," that is, a characteristic curve is the level curve of the function r, the vector $\text{grad}\, r$ is orthogonal to this characteristic line (see Fig. 4.2):

$$\text{grad}\, r \perp (dx, dt).$$

Therefore, $\text{grad}\, r \| (dt, -dx)$, so that $\frac{\frac{\partial r}{\partial x}}{\frac{\partial r}{\partial t}} = -\frac{dt}{dx}$, or

$$dt = -k\, dx, \quad \text{where} \quad k = \frac{\frac{\partial r}{\partial x}}{\frac{\partial r}{\partial t}}. \tag{4.21}$$

Substituting (4.21) into (4.17), we obtain the desired differential equation:

$$a(x,t)\left(\frac{\partial r}{\partial t}\right)^2 + 2b(x,t)\frac{\partial r}{\partial t}\frac{\partial r}{\partial x} + c(x,t)\left(\frac{\partial r}{\partial x}\right)^2 = 0. \tag{4.22}$$

In the same fashion one derives the differential equation for $s(x,t)$, and it coincides with (4.22):

$$a(x,t)\left(\frac{\partial s}{\partial t}\right)^2 + 2b(x,t)\frac{\partial s}{\partial t}\frac{\partial s}{\partial x} + c(x,t)\left(\frac{\partial s}{\partial x}\right)^2 = 0. \tag{4.23}$$

This is of no surprise, since (4.17) contains both equations from (4.20).

Now let us recall formulas (2.8)–(2.9) for the change of variables in a differential equation. Substituting expressions (2.8)–(2.9) into (4.16), we obtain the following differential equation for the function $v(r,s) = u(x,t)$:

$$\alpha(r,s)\frac{\partial^2 v}{\partial r^2} + 2\beta(r,s)\frac{\partial^2 v}{\partial r \partial s} + \gamma(r,s)\frac{\partial^2 v}{\partial s^2} + \ldots = 0,$$

where "$\ldots$" stands for the terms containing first order derivatives of v. The expressions for the coefficients α, β are γ are as follows:

$$\alpha = a\left(\frac{\partial r}{\partial t}\right)^2 + 2b\frac{\partial r}{\partial t}\frac{\partial r}{\partial x} + c\left(\frac{\partial r}{\partial x}\right)^2, \tag{4.24}$$

$$\gamma = a\left(\frac{\partial s}{\partial t}\right)^2 + 2b\frac{\partial s}{\partial t}\frac{\partial s}{\partial x} + c\left(\frac{\partial s}{\partial x}\right)^2, \tag{4.25}$$

$$\beta = a\frac{\partial r}{\partial t}\frac{\partial s}{\partial t} + 2b\left(\frac{\partial r}{\partial t}\frac{\partial s}{\partial x} + \frac{\partial s}{\partial t}\frac{\partial r}{\partial x}\right) + c\frac{\partial r}{\partial x}\frac{\partial s}{\partial x}. \tag{4.26}$$

Denote by A the *characteristic polynomial* of equation (4.16) at the point (x,t):

$$A(x,t;\xi,\tau) \equiv a(x,t)\tau^2 + 2b(x,t)\tau\xi + c(x,t)\xi^2. \tag{4.27}$$

Then (4.22) and (4.23) are equivalent to

$$\alpha = A(\operatorname{grad} r(x,t)) = 0, \qquad \gamma = A(\operatorname{grad} s(x,t)) = 0. \tag{4.28}$$

Finally, (4.16) takes the form similar to (4.11):

$$2\beta(r,s)\frac{\partial^2 v}{\partial r \partial s} + \ldots = 0. \tag{4.29}$$

Problem 4.4. Prove that $\beta(r,s) \neq 0$ at $r = r(x,t)$, $s = s(x,t)$ if the condition (4.19) holds at the point (x,t).

Hint. Use (4.24)–(4.28).

Equation (4.29) could be solved approximately. In a number of cases, when equation (4.29) is sufficiently simple, it is possible to find its general solution and thus to find the general solution to equation (4.16).

Problem 4.5. Find the general solution to the equation

$$\frac{\partial^2 u}{\partial x^2} - 2\sin x\frac{\partial^2 u}{\partial x \partial y} - \cos^2 x\frac{\partial^2 u}{\partial y^2} - \cos x\frac{\partial u}{\partial y} = 0. \tag{4.30}$$

Solution. The characteristic equation (4.17) is obtained from (4.30) by substituting $\frac{\partial}{\partial x} \mapsto dy$; $\frac{\partial}{\partial y} \mapsto -dx$ (see (4.18)):

$$dy^2 + 2\sin x\, dy\, dx - \cos^2 x\, dx^2 = 0$$

or
$$\left(\frac{dy}{dx}\right)^2 + 2\sin x \frac{dy}{dx} - \cos^2 x = 0.$$

From here,
$$\frac{dy}{dx} = -\sin x \pm \sqrt{\sin^2 x + \cos^2 x} = -\sin x \pm 1. \tag{4.31}$$

Integrating, we get
$$y = \cos x \pm x = c.$$

Hence the functions
$$c(x,y) = y - \cos x \mp x$$

are constant along the integral curves, that is, represent the first integrals of equation (4.31). Therefore,
$$\begin{cases} r = y - \cos x - x, \\ s = y - \cos x + x. \end{cases} \tag{4.32}$$

We already know that equation (4.30) in the variables r, s has the form (4.29). But we also need to know the form of the terms containing $\frac{\partial v}{\partial r}$, $\frac{\partial v}{\partial s}$, which are not written explicitly in (4.29). We could use the known formulas (2.8), (2.9), but let us make the change of variables (4.32) in (4.30) directly. Writing u instead of v, we have:
$$\begin{cases} \frac{\partial u}{\partial x} = \frac{\partial u}{\partial r}\frac{\partial r}{\partial x} + \frac{\partial u}{\partial s}\frac{\partial s}{\partial x} = \frac{\partial u}{\partial r}(\sin x - 1) + \frac{\partial u}{\partial s}(\sin x + 1), \\ \frac{\partial u}{\partial y} = \frac{\partial u}{\partial r} + \frac{\partial u}{\partial s}. \end{cases} \tag{4.33}$$

We then have
$$\frac{\partial^2 u}{\partial x^2} = \ldots + 2\frac{\partial^2 u}{\partial r \partial s}(\sin^2 x - 1) + \ldots + \frac{\partial u}{\partial r}\cos x + \frac{\partial u}{\partial s}\cos x.$$

Dots denote the terms containing $\frac{\partial^2 u}{\partial r^2}$ and $\frac{\partial^2 u}{\partial s^2}$, which, as we already know (see (4.29)), cancel out in (4.30). Therefore, we do not have to write them out!

Analogously,
$$\frac{\partial^2 u}{\partial x \partial y} = \ldots + \frac{\partial^2 u}{\partial r \partial s}(\sin x - 1) + \frac{\partial^2 u}{\partial s \partial r}(\sin x + 1) + \ldots.$$

Finally,
$$\frac{\partial^2 u}{\partial y^2} = \ldots + 2\frac{\partial^2 u}{\partial r \partial s} + \ldots. \tag{4.34}$$

Substituting (4.33)–(4.34) into (4.30), we get
$$\frac{\partial^2 u}{\partial r \partial s}\left(2(\sin^2 x - 1) - 2\sin x \cdot 2\sin x - 2\cos^2 x\right)$$
$$+ \left(\frac{\partial u}{\partial r} + \frac{\partial u}{\partial s}\right)\cos x - \left(\frac{\partial u}{\partial r} + \frac{\partial u}{\partial s}\right)\cos x = 0.$$

4 Second-order hyperbolic equations in the plane

After cancellations and collecting the terms, we obtain

$$\frac{\partial^2 u}{\partial r \partial s} = 0, \quad \text{hence} \quad u = f(r) + g(s).$$

Answer.

$$u(x,y) = f(y - \cos x - x) + g(y - \cos x + x).$$

For better understanding of the material we recommend to solve the following problems:

Problem 4.6. Find the general solution for the following equations:

a. $\frac{\partial^2 u}{\partial x^2} + 2\frac{\partial^2 u}{\partial x \partial y} - 3\frac{\partial^2 u}{\partial y^2} + 2\frac{\partial u}{\partial x} + 6\frac{\partial u}{\partial y} = 0$.

b. $x\frac{\partial^2 u}{\partial x^2} - y\frac{\partial^2 u}{\partial y^2} + \frac{1}{2}\left(\frac{\partial u}{\partial x} - \frac{\partial u}{\partial y}\right) = 0, \quad x > 0, \ y > 0$.

c. $x^2\frac{\partial^2 u}{\partial x^2} - y^2\frac{\partial^2 u}{\partial y^2} - 2y\frac{\partial u}{\partial y} = 0$.

d. $\frac{\partial}{\partial x}\left(x^2\frac{\partial u}{\partial x}\right) = x^2\frac{\partial^2 u}{\partial y^2}$.

e. $(x - y)\frac{\partial^2 u}{\partial x \partial y} - \frac{\partial u}{\partial x} + \frac{\partial u}{\partial y} = 0$.

f. $\frac{\partial^2 u}{\partial x \partial y} + y\frac{\partial u}{\partial x} + x\frac{\partial u}{\partial y} + xyu = 0$.

Problem 4.7. Solve the following Cauchy problems:

a. $\frac{\partial^2 u}{\partial x^2} + 2\frac{\partial^2 u}{\partial x \partial y} - 3\frac{\partial^2 u}{\partial y^2} = 0, \quad u\big|_{y=0} = 3x^2, \ \frac{\partial u}{\partial y}\big|_{y=0} = 0$.

b. $4y^2\frac{\partial^2 u}{\partial x^2} + 2(1 - y^2)\frac{\partial^2 u}{\partial x \partial y} - \frac{\partial^2 u}{\partial y^2} - \frac{2y}{1+y^2}\left(2\frac{\partial u}{\partial x} - \frac{\partial u}{\partial y}\right) = 0,$
$\quad u\big|_{y=0} = \varphi_0(x), \quad \frac{\partial u}{\partial y}\big|_{y=0} = \varphi_1(x)$.

c. $(1 + x^2)\frac{\partial^2 u}{\partial x^2} - (1 + y^2)\frac{\partial^2 u}{\partial y^2} + x\frac{\partial u}{\partial x} - y\frac{\partial u}{\partial y} = 0, \quad u\big|_{y=0} = \varphi_0(x), \quad \frac{\partial u}{\partial y}\big|_{y=0} = \varphi_1(x)$.

d. $\frac{\partial^2 u}{\partial x^2} + 2\cos x\frac{\partial^2 u}{\partial x \partial y} - \sin^2 x\frac{\partial^2 u}{\partial y^2} - \sin x\frac{\partial u}{\partial y} = 0, \quad u\big|_{y=\sin x} = \varphi_0(x), \quad \frac{\partial u}{\partial y}\big|_{y=\sin x} = \varphi_1(x)$.

e. $\frac{\partial^2 u}{\partial x^2} + 4\frac{\partial^2 u}{\partial x \partial y} - 5\frac{\partial^2 u}{\partial y^2} + \frac{\partial u}{\partial x} - \frac{\partial u}{\partial y} = 0, \quad u\big|_{y=0} = f(x), \quad \frac{\partial u}{\partial y}\big|_{y=0} = F(x)$.

f. $x^2\frac{\partial^2 u}{\partial x^2} - 2xy\frac{\partial^2 u}{\partial x \partial y} - 3y^2\frac{\partial^2 u}{\partial y^2} = 0, \quad u\big|_{y=1} = \varphi_0(x), \quad \frac{\partial u}{\partial y}\big|_{y=1} = \varphi_1(x)$.

Nonhyperbolic equations

Let us consider the case, when instead of the strict hyperbolicity condition (4.19) the opposite inequality holds:

$$b^2(x,t) - a(x,t)c(x,t) < 0.$$

In this case, equation (4.16) is called elliptic at the point (x,t). The right-hand side of equations (4.20) are complex conjugates, and the integration yields the "first integrals" r and $s = \bar{r}$ which are also complex conjugates. It turns out that if one takes $z_1 = \operatorname{Re} r = \frac{r+s}{2}$ and $z_2 = \operatorname{Im} r = \frac{r-s}{2i}$ as the new coordinates, then equation (4.16) takes the form

$$\frac{\partial^2 u}{\partial z_1^2} + \frac{\partial^2 u}{\partial z_2^2} + \ldots = 0, \tag{4.35}$$

that is, its principal part coincides with the Laplace operator. This allows us to solve such equations exactly or approximately.

Problem 4.8. Bring the following equation to the canonical form:

$$y\frac{\partial^2 u}{\partial x^2} + x\frac{\partial^2 u}{\partial y^2} = 0, \qquad y > 0, \quad x > 0. \tag{4.36}$$

Solution. The equation of the characteristics $y\,dy^2 + x\,dx^2 = 0$ takes the form $\sqrt{y}\,dy = \pm i\sqrt{x}\,dx$, that is, equation (4.36) is elliptic. Integrating, we get $y^{3/2} \mp ix^{3/2} = c$. Take the new coordinates $z_1 = \operatorname{Re} c = y^{3/2}$, $z_2 = \operatorname{Im} c = x^{3/2}$. Then

$$\frac{\partial u}{\partial x} = \frac{\partial u}{\partial z_2}\frac{3}{2}x^{1/2}, \qquad \frac{\partial u}{\partial y} = \frac{\partial u}{\partial z_1}\frac{3}{2}y^{1/2}.$$

Differentiating the above relations in x and y, respectively, we get (in accordance with (4.35), we do not write the terms with $\frac{\partial^2 u}{\partial z_1 \partial z_2}$):

$$\frac{\partial^2 u}{\partial x^2} = \frac{\partial^2 u}{\partial z_2^2}\frac{9}{4}x + \frac{\partial u}{\partial z_2}\frac{3}{4}x^{-1/2} + \ldots, \qquad \frac{\partial^2 u}{\partial y^2} = \frac{\partial^2 u}{\partial z_1^2}\frac{9}{4}y + \frac{\partial u}{\partial z_1}\frac{3}{4}y^{-1/2} + \ldots.$$

Substituting this into (4.36), we find

$$\left(\frac{\partial^2 u}{\partial z_1^2} + \frac{\partial^2 u}{\partial z_2^2}\right)\frac{9}{4}xy + \frac{3}{4}\frac{\partial u}{\partial z_1}xy^{-1/2} + \frac{3}{4}\frac{\partial u}{\partial z_2}yx^{-1/2} = 0.$$

From here we get the canonical form:

$$\frac{\partial^2 u}{\partial z_1^2} + \frac{\partial^2 u}{\partial z_2^2} + \frac{1}{3z_1}\frac{\partial u}{\partial z_1} + \frac{1}{3z_2}\frac{\partial u}{\partial z_2} = 0.$$

Now let us consider the case, when in (4.19) instead of ">" one has "=." Then equation (4.16) is called degenerate at the point (x,t). If (4.16) is degenerate in a certain region, then equations (4.20) coincide and consequently there is only one independent first integral $r(x,t)$. In this case, for bringing (4.16) to the canonical form, one could choose as a second variable any function so that the change of variables $(x,t) \mapsto (r,s)$ were non-degenerate. It turns out that (4.16) takes the form

$$\frac{\partial^2 u}{\partial s^2} + \ldots = 0. \tag{4.37}$$

Problem 4.9. Bring to the canonical form the equation

$$\sin^2 x \frac{\partial^2 u}{\partial x^2} - 2y \sin x \frac{\partial^2 u}{\partial x \partial y} + y^2 \frac{\partial^2 u}{\partial y^2} = 0, \qquad 0 < x < \pi, \quad y > 0. \tag{4.38}$$

Solution. The equation of the characteristics, $\sin^2 x \, dy^2 + 2y \sin x \, dy \, dx + y^2 \, dx^2 = 0$ takes the form $(\sin x \, dy + y \, dx)^2 = 0$, that is, equation (4.38) is degenerate. Separating the variables, we get $dx/\sin x = -dy/y$, hence $\ln \tan \frac{x}{2} = -\ln y + c$, or $y \tan \frac{x}{2} = c_1$. We take $r = y \tan \frac{x}{2}$. Then, setting $s = y$, we find

$$\frac{\partial u}{\partial x} = \frac{\partial u}{\partial r} y \frac{1}{2\cos^2 \frac{x}{2}}, \qquad \frac{\partial u}{\partial y} = \frac{\partial u}{\partial r} \tan \frac{x}{2} + \frac{\partial u}{\partial s}. \tag{4.39}$$

Differentiating, we get (omitting the terms with $\frac{\partial^2 u}{\partial r \partial s}$ and $\frac{\partial^2 u}{\partial r^2}$, in accordance with (4.37)):

$$\frac{\partial^2 u}{\partial x^2} = \frac{\partial u}{\partial r} y \frac{\sin \frac{x}{2}}{2\cos^3 \frac{x}{2}} + \ldots, \qquad \frac{\partial^2 u}{\partial y^2} = \frac{\partial^2 u}{\partial s^2} + \ldots, \qquad \frac{\partial^2 u}{\partial x \partial y} = \frac{\partial u}{\partial r} \frac{1}{2\cos^2 \frac{x}{2}} + \ldots.$$

Substituting into (4.38), we find

$$\sin^2 x \left(\frac{\partial u}{\partial r} y \frac{\sin \frac{x}{2}}{2\cos^3 \frac{x}{2}} \right) - 2y \sin x \frac{\partial u}{\partial r} \frac{1}{2\cos^2 \frac{x}{2}} + y^2 \frac{\partial^2 u}{\partial s^2} = 0,$$

from where we obtain the canonical form:

$$\frac{\partial^2 u}{\partial s^2} + \frac{\partial u}{\partial r} \left(\frac{r \sin^2 x}{y^2 2\cos^2 \frac{x}{2}} - \frac{\sin x}{y \cos^2 \frac{x}{2}} \right) = \frac{\partial^2 u}{\partial s^2} + \frac{\partial u}{\partial r} \left(-\frac{2r}{s^2 + r^2} \right) = 0.$$

Problem 4.10. Bring to the canonical form the following equations:

a. $\frac{\partial^2 u}{\partial x^2} - 2\frac{\partial^2 u}{\partial x \partial y} + \frac{\partial^2 u}{\partial y^2} + \alpha \frac{\partial u}{\partial x} + \beta \frac{\partial u}{\partial y} + cu = 0.$

b. $\tan^2 x \frac{\partial^2 u}{\partial x^2} - 2y \tan x \frac{\partial^2 u}{\partial x \partial y} + y^2 \frac{\partial^2 u}{\partial y^2} + \tan^3 x \frac{\partial u}{\partial x} = 0.$

c. $\coth^2 x \frac{\partial^2 u}{\partial x^2} - 2y \coth x \frac{\partial^2 u}{\partial x \partial y} + y^2 \frac{\partial^2 u}{\partial y^2} + 2y \frac{\partial u}{\partial y} = 0.$

5 Semi-infinite string

Mixed problem for the d'Alembert equation

Let us consider the d'Alembert equation (2.1) in the region $x > 0$. Physically, this corresponds to the string with one (left) end located at the origin and the other located far away from the origin (at a distance $\gg at$):

$$\frac{\partial^2 u}{\partial t^2} = a^2 \frac{\partial^2 u}{\partial x^2}, \qquad x > 0, \qquad t > 0. \tag{5.1}$$

The initial conditions (2.2) are also required here:

$$u(x,0) = \varphi(x), \qquad \frac{\partial u}{\partial t}(x,0) = \psi(x), \qquad x > 0. \tag{5.2}$$

Besides, it is physically obvious that one needs the boundary condition at the left end of the string (at $x = 0$). For example, if this end is fixed, then its displacement is equal to zero:

$$u(0,t) = 0, \qquad t > 0. \tag{5.3}$$

Other physically sensible boundary conditions are also possible; see, for example, (1.12) and (1.18).

The problem (5.1)–(5.3) is called a *mixed problem*, since it contains both the initial data (5.2) and the boundary conditions (5.3).

Solution of the mixed problem. Incident and reflected waves

Let us use the d'Alembert method, that is, let us search for a solution in the form

$$u(x,t) = f(x - at) + g(x + at). \tag{5.4}$$

Substituting this decomposition into the initial data (5.2), we get, as in Section 2, equations (2.17)–(2.20), that is, the d'Alembert formula for $u(x,t)$.

Question 5.1. Why do we need the boundary condition (5.3), if we seem to have found the solution using only the initial data?

Answer. Equations (2.17)–(2.19) only make sense for $x > 0$, since the initial data (5.2), as opposed to (2.2), are only given for $x > 0$. Correspondingly, the d'Alembert formula (2.20) only holds for $x - at > 0$, and not for all $x > 0, t > 0$.

Conclusion. The solution to the mixed problem (5.1)–(5.3) is given by the d'Alembert formula (2.20) for $x - at > 0$.

This is the region below the *principal characteristic curve* $x - at = 0$ (Fig. 5.1).

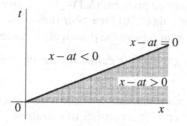

Fig. 5.1

The characteristic curve $x - at = 0$ is called *principal* since it comes out of a special point (corner point) of the region $x > 0, t > 0$ where equation (5.1) is considered. Now let us find the solution above the principal characteristic curve (in the region $x - at < 0$). Decomposition (5.4) holds everywhere in the region $x > 0, t > 0$. The wave $g(x + at)$ is found from (2.18) for all $x > 0, t > 0$. On the other hand, the wave $f(x - at)$ is found from (2.19) only in the region $x - at > 0$, that is, below the principal characteristic curve. Thus, it remains to find $f(x - at)$ above the principal characteristic curve, for $x - at < 0$.

Let us find $f(x - at)$ for $x - at < 0$. We use the boundary condition (5.3):

$$f(-at) + g(at) = 0, \qquad t > 0. \tag{5.5}$$

It is this formula that relates the unknown values of the function f for the negative values of its argument with the values of the function g for the positive values of its argument, which are already known from (2.18).

Let us make the change of variables: We set $-at = z$. Then (5.5) takes the form

$$f(z) = -g(-z), \qquad z < 0.$$

Due to (2.18), the above relation shows that for $x - at < 0$

$$f(x - at) = -g(at - x) = -\frac{\varphi(at - x)}{2} - \frac{1}{2a}\int_0^{at-x} \psi(y)\,dy - \frac{c}{2a}$$

$$= -\frac{\varphi(at - x)}{2} + \frac{1}{2a}\int_{at-x}^0 \psi(y)\,dy - \frac{c}{2a}. \tag{5.6}$$

Substituting (5.6) and (2.18) into (5.4), we find out the following: for $x > at$, the solution is given by the d'Alembert formula (2.20); for $0 < x < at$, the solution is given by

$$u(x,t) = \frac{-\varphi(at - x) + \varphi(x + at)}{2} + \frac{1}{2a}\int_{at-x}^{x+at} \psi(y)\,dy. \tag{5.7}$$

Thus, the solution of the mixed problem (5.1)–(5.3) is given by two different formulas: The d'Alembert formula (2.20) for $x > at$ (below the principal characteristic curve) and (5.7) for $0 < x < at$ (above the principal characteristic curve).

Definition 5.2. *The wave $g(x + at)$ is called the* incident wave, *while $f(x - at)$ is called the* reflected wave.

Let us give *the graphical interpretation* of constructing a solution to the problem (5.1)–(5.3). Solution of this problem consists of two steps:
A. We substitute the d'Alembert decomposition (5.4) into the *initial data* (5.2), which are specified at $t = 0$ at the points $x > 0$ of the Ox axis.

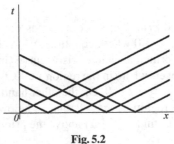

Fig. 5.2

Solving the system (2.17) for $x > 0$, we find the waves $f(x - at)$ and $g(x + at)$ at these same points $x > 0$, $t = 0$. Now $f(x - at)$ is known on all characteristics going to the right from these points (Fig. 5.2) since $f(x - at)$ is constant on all such characteristics. These characteristics fill the entire region $x - at > 0$. On the other hand, the wave $g(x + at)$ is known everywhere. Indeed, it is constant on the characteristics going to the left, while such characteristics, sent out of the points (x, t) with $x > 0$ and $t = 0$, fill the entire region $x > 0$, $t > 0$. Thus, the initial data allow to determine the solution in the region on Fig. 5.2 containing the characteristics of both families, that is, below the principal characteristic curves.

One can see on Fig. 5.2 that above the principal characteristic curve the wave $f(x - at)$ (the reflected wave) is not known yet, while the incident wave $g(x + at)$ is already known.
B. We substitute the d'Alembert decomposition (5.4) into the boundary condition (5.3), which is specified at the points of the time axis Ot ($t > 0$, $x = 0$). At these points the wave $g(x + at)$ is already determined from the initial data. Therefore the boundary condition (5.5) relates the values of the wave $f(x - at)$ (unknown at these points) with the already known values of $g(x + at)$. This allows us to determine the wave $f(x - at)$. But then $f(x - at)$ (and

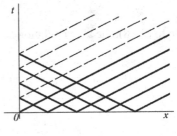

Fig. 5.3

hence $u(x, t)$) is known on the characteristics going to the right from all these points (the dashed line on Fig. 5.3), that is, in the entire region $x < at$ above the principal characteristic curve.

Other boundary conditions

Instead of the boundary condition (5.3), one may consider the boundary condition (1.10):

$$\frac{\partial u}{\partial x}(0,t) = 0, \qquad t > 0. \tag{5.8}$$

Problem 5.3. Solve the mixed problem (5.1)–(5.2), (5.8).

Solution.

a. Below the principal characteristic curve, that is, for $x > at$, the d'Alembert formula (2.20) is valid, and formulas (2.18)–(2.19) hold for $x > 0$;

b. Above the principal characteristic curves, that is, for $x < at$, we substitute (5.4) into the boundary condition (5.8), obtaining

$$f'(-at) + g'(at) = 0, \qquad t > 0. \tag{5.9}$$

After the substitution $-at = z$, we have:

$$f'(z) + g'(-z) = 0, \qquad z < 0.$$

Integrating, we get

$$f(z) - g(-z) = c_1 = \text{const}, \qquad z < 0.$$

In view of (2.18), we obtain the solution for $x < at$:

$$f(x - at) = g(at - x) + c_1 = \frac{1}{2}\varphi(at - x) + \frac{1}{2a} \int\limits_0^{at-x} \psi(y)\,dy + \frac{c}{2a} + c_1. \tag{5.10}$$

Taking $g(x + at)$ from the formula (2.18), we also obtain the solution for $x < at$:

$$u(x,t) = \frac{\varphi(at - x) + \varphi(x + at)}{2} + \frac{1}{2a} \int\limits_0^{at-x} \psi(y)\,dy + \frac{1}{2a} \int\limits_0^{x+at} \psi(y)\,dy + c_2. \tag{5.11}$$

The constant c_2, as we will now show, could be determined from the condition that the solution $u(x,t)$ is continuous at the characteristic curve $x = at$. The continuity is necessary when the problem (5.1)–(5.2), (5.8) describes a string or a rod.

Discontinuities of a solution along a principal characteristic curve. Continuity conditions

It follows that the solution to the problem (5.1)–(5.2) is given by different expressions for $x - at > 0$ and $x - at < 0$, therefore it could be discontinuous along the line $x - at = 0$. It turns out that the discontinuity of any solution to (5.1) along the line $x - at = 0$ does not depend on time.

Indeed, this could be seen from (5.4):

a. The wave $g(x + at)$ is continuous when passing through the principal characteristic curve, since its level curves $x + at = \text{const}$ intersect the line $x = at$.

b. The wave $f(x - at)$ below the principal characteristic curve $x - at = 0$ has a limit equal to $f(0+)$, since in that region one has $x - at > 0$; analogously, its limit from above is equal to $f(0-)$. Thus,

$$u\big|_{x-at=0-} - u\big|_{x-at=0+} = f(0-) - f(0+).$$

Therefore, the condition that the solution $u(x,t)$ is continuous on the principal characteristic curve has the form

$$f(0-) = f(0+). \tag{5.12}$$

Remark 5.4. We use the notations

$$f(a\pm) := \lim_{x \to a \pm 0} f(x).$$

Problem 5.5. Find the condition for the solution to the problem (5.1)–(5.3) to be continuous at the principal characteristic curve.

Solution. As it follows from (2.19),

$$f(0+) = \frac{\varphi(0)}{2} - \frac{c}{2a}, \tag{5.13}$$

while from (5.6) we have

$$f(0-) = -g(0) = -\frac{\varphi(0)}{2} - \frac{c}{2a}. \tag{5.14}$$

Therefore, the condition (5.12) gives

$$-\frac{\varphi(0)}{2} = \frac{\varphi(0)}{2}, \quad \text{hence} \quad \varphi(0) = 0. \tag{5.15}$$

Remark 5.6. Let us consider the region $x > 0, t > 0$ (see Fig. 5.4) where the problem (5.1)–(5.3) is being solved. On the part of its boundary represented by the axis Ot, the solution is equal to zero due to (5.3), while at the axis Ox the solution is equal to $\varphi(x)$. Therefore the condition (5.15) is merely the condition for the boundary

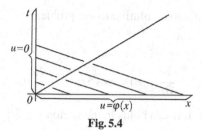

Fig. 5.4

values of $u(x,t)$ to be continuous at the point $(0,0)$. As we have seen, this condition is necessary and sufficient for the continuity of the solution at all the points of the principal characteristic curve.

Problem 5.7. Find the condition for the solution to the problem (5.1)–(5.2), (5.8) to be continuous at the principal characteristic curve.

Solution. The formula (5.13) for $f(0+)$ from the previous problem is valid here. The expression

$$f(0-) = \frac{\varphi(0)}{2} + \frac{c}{2a} + c_1$$

follows from (5.10). Therefore, (5.11) and (5.12) imply that

$$\frac{\varphi(0)}{2} + \frac{c}{2a} + c_1 = \frac{\varphi(0)}{2} - \frac{c}{2a}, \quad \text{hence} \quad c_1 + \frac{c}{a} = c_2 = 0. \quad (5.16)$$

Remark 5.8. A discontinuous solution to the problem (5.1)–(5.2), (5.8) (when $c_2 \neq 0$) does not make a physical sense for a string or a rod, since it implies their breaking. Yet, in acoustics and gas dynamics a discontinuous solution makes physical sense and is called a shock wave. In this case, the value of discontinuity, represented by c_2, could not be found from equations (5.1), (5.2), and (5.8).

This value can be determined from additional physical or chemical information, allowing us to pinpoint a unique solution to the problem. For example, in the process of propagation of the detonation wave in the gasoline vapor the value of the pressure jump at the front of the shock wave depends on the type of the gasoline, pressure, temperature, presence of additional substances, and so on.

The mixed problem (5.1)–(5.2) with more general boundary conditions (1.12) or (1.18) is solved similarly as in the case of the boundary condition (5.8), but the equation of type (5.9) for the boundary condition for the reflected wave will be the second order differential equation, and its solution will contain two arbitrary constants. These constants are determined in each particular problem from the auxiliary conditions. For example, the condition (5.21) below means that the mass at $t = 0$ is attached to the left end of the rod and its (horizontal) speed is equal to 7.

Problem 5.9. Find a continuous solution to the problem

$$
\begin{cases}
\dfrac{\partial^2 u}{\partial t^2} = 9\dfrac{\partial^2 u}{\partial x^2}, & x > 0, \quad t > 0; \\
u(x,0) = e^{-x}, & \dfrac{\partial u}{\partial t}(x,0) = \cos 5x; \qquad \dfrac{\partial u}{\partial x}(0,t) = u(0,t) + t.
\end{cases}
$$

Solution. The d'Alembert formula holds in the region $x > 3t$:

$$
u(x,t) = \frac{e^{-(x-3t)} + e^{-(x+3t)}}{2} + \frac{1}{6}\frac{\sin\left(5(x+3t)\right) - \sin\left(5(x-3t)\right)}{5}. \tag{5.17}
$$

Therefore, for $0 < x < 3t$, one needs to look for a solution in the form

$$
u(x,t) = f(x - 3t) + \frac{e^{-(x+3t)}}{2} + \frac{\sin\left(5(x+3t)\right)}{30}.
$$

Substituting this expression into the boundary condition, we find:

$$
f'(-3t) - \frac{e^{3t}}{2} + \frac{1}{6}\cos 15t = f(-3t) + \frac{e^{-3t}}{2} + \frac{\sin 15t}{30} + t, \qquad t > 0.
$$

Substituting $y = -3t$, we obtain:

$$
f'(y) - \frac{e^{y}}{2} + \frac{1}{6}\cos 5y = f(y) + \frac{e^{y}}{2} - \frac{\sin 5y}{30} - \frac{y}{3}, \qquad y < 0,
$$

or

$$
f'(y) - f(y) = e^{y} - \frac{1}{6}\cos 5y - \frac{\sin 5y}{30} - \frac{y}{3}, \qquad y < 0. \tag{5.18}
$$

It follows that

$$
f(y) = Ce^{y} + ye^{y} + A\cos 5y + B\sin 5y + \frac{y}{3} + \frac{1}{3}, \qquad y < 0.
$$

We find the values of constants A and B substituting $f(y)$ into (5.18):

$$
-5A\sin 5y - A\cos 5y + 5B\cos 5y - B\sin 5y = -\frac{\cos 5y}{6} - \frac{\sin 5y}{30}.
$$

Therefore $-5A - B = \frac{1}{30}$; $-A + 5B = -\frac{1}{6}$, and thus $-26A = -\frac{1}{3}$, leading to $A = \frac{1}{78}$ and $B = -5A + \frac{1}{30} = -\frac{5}{78} + \frac{1}{30}$. Finally, C could be found from the continuity condition (5.12): $C + A + \frac{1}{3} = \frac{1}{2}$, hence $C = \frac{1}{6} - A = \frac{1}{6} - \frac{1}{78} = \frac{2}{13}$.

Answer. For $x > 3t$, $u(x,t)$ is given by (5.17). For $0 < x < 3t$,

$$
u(x,t) = \frac{2}{13}e^{x-3t} + (x - 3t)e^{x-3t} + \frac{1}{78}\cos 5(x - 3t)
$$
$$
+ \left(\frac{1}{30} - \frac{5}{78}\right)\sin 5(x - 3t) + \frac{1}{3}(x - 3t) + \frac{1}{3} + \frac{1}{2}e^{-(x+3t)} + \frac{1}{30}\sin 5(x + 3t).
$$

Propagation of waves

Problem 5.10. A stretched semi-infinite rope described by (5.1) with $a = 1$ is initially at rest. Starting at $t = 0$, its left end $x = 0$ is moved up and down, with the displacement being equal to $\sin \pi t$. Draw the shape of the rope at $t = 1, 2, 3, \ldots$.

Solution. We need to solve the mixed problem (5.1)–(5.2) with $\varphi(x) \equiv \psi(x) \equiv 0$ and with the boundary condition

$$u(0,t) = \sin \pi t, \qquad t > 0. \tag{5.19}$$

(*i*) $x > t$: in this case, since $\varphi(x) \equiv \psi(x) \equiv 0$ by the condition of the problem, $u(x,t) = 0$. In particular, $g(x+t) \equiv 0$ for all $x > 0, t > 0$.
(*ii*) $x < t$: since $g(x+t) \equiv 0$, $u(x,t) \equiv f(x-t)$. Substituting $u(x,t) = f(x-t)$ into (5.19), we get $f(-t) = \sin \pi t$, $t > 0$.
Answer. $u(x,t) = f(x-t) = \sin \pi(t-x) = -\sin \pi(x-t)$, $x < t$. See Fig. 5.5.

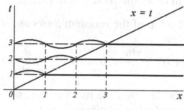

Fig. 5.5

Problem 5.11. A stretched rope described by (5.1) with $a = 1$ is initially at rest. Starting at $t = 0$ its left end $x = 0$ is moved up and down by the external force $f(t) = \sin \pi t$ (see the boundary condition (1.12), where we take $T = 1$, $m = k = \eta = 0$). Draw the shape of the string at $t = 1, 2, 3, \ldots$.

Solution. We need to find a continuous solution to the mixed problem (5.1)–(5.2) with $\varphi(x) \equiv \psi(x) \equiv 0$ and with the boundary condition

$$\frac{\partial u}{\partial x}(0,t) = -\sin \pi t, \qquad t > 0. \tag{5.20}$$

a. $x > t$: in this case, $u(x,t) \equiv 0$. In particular, $g(x+t) \equiv 0$.
b. $x < t$: in this case, $u(x,t) = f(x-t)$.

Substituting $u(x,t) = f(x-t)$ into (5.20), we get $f'(-t) = -\sin \pi t$ for $t > 0$. Substituting $-t = z$, we can write $f'(z) = \sin \pi z$ for $z < 0$, hence $f(z) = -\frac{\cos \pi z}{\pi} + c$, $z < 0$. Therefore,

$$u(x,t) = f(x-t) = -\frac{\cos \pi(x-t)}{\pi} + c, \qquad x < t.$$

The continuity condition at $x = t$ requires that $u(t,t) = 0 = -\frac{1}{\pi} + c$, hence $c = \frac{1}{\pi}$.

Answer. $u(x,t) = \left(1 - \cos \pi(x-t)\right)/\pi$ for $x < t$ and zero otherwise. See Fig. 5.6.

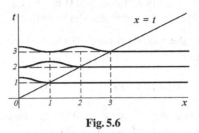

Fig. 5.6

Problem 5.12. The mass $m = 2$ moving with the speed $v = 7$ sticks to the end of the semi-infinite rod that was initially at rest. Find the displacement of the rod for $t > 0$, assuming that $a = 3$ in (5.1) and $S \cdot E = 5$ in (1.18).

Solution. The mathematical setup of the problem looks as follows:

$$\frac{\partial^2 u}{\partial t^2} = 9 \frac{\partial^2 u}{\partial x^2}, \qquad u(x,0) = \frac{\partial u}{\partial t}(x,0) = 0, \qquad 2 \frac{\partial^2 u}{\partial t^2}(0,t) = 5 \frac{\partial u}{\partial x}(0,t).$$

The last relation is due to the fact that the mass at the end of the rod is due to the newly acquired mass m. The sticking of the mass to the end of the rod leads to the following conditions:

$$u(0,0+) = 0, \qquad \frac{\partial u}{\partial t}(0,0+) = 7. \qquad (5.21)$$

For $x > 3t$ the d'Alembert formula holds, so that $u(x,t) = 0$, since the initial data are equal to zero. For $x < 3t$, since $g(x+3t) \equiv 0$, the solution has the form

$$u(x,t) = f(x-3t).$$

Substituting $u(x,t)$ in this form into the boundary conditions, we find:

$$2 \cdot 9 f''(-3t) = 5 f'(-3t), \quad t > 0; \qquad f(0-) = 0; \qquad -3 f'(0-) = 7.$$

Hence $18 f''(y) - 5 f'(y) = 0$ for $y < 0$, giving $f(y) = c_1 + c_2 e^{\frac{5}{18}y}$; $c_1 + c_2 = 0$. Evaluating $f'(0-)$, we find that $-3 c_2 \frac{5}{18} = 7$, $c_2 = -\frac{42}{5}$.

Answer. $u = 0$ for $x > 3t$ and $u = \frac{42}{5}\left(1 - e^{\frac{5}{18}(x-3t)}\right)$ for $x < 3t$.

Reflection of waves. Method of even and odd extension

Besides the general method described above, the problem (5.1), (5.2) with the boundary conditions (5.3) or (5.8) could be approached using the method of odd and even extension.

Let us first consider *the method of odd extension*. The following problem describes oscillations of a plucked string.

Problem 5.13. Solve the mixed problem (5.1)–(5.3) with $a = 1$ and the initial data from Fig. 5.7. Draw the shape of the string at $t = 1, 2, 3, 3.5, 4, 5$.

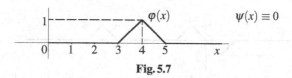

Fig. 5.7

Solution. Let us consider the solution $\hat{u}(x,t)$ to the Cauchy problem (2.1)–(2.2) on the entire axis, with $\frac{\partial}{\partial t}\hat{u}(x,0) = \hat{\psi}(x) \equiv 0$ and with $\hat{\varphi}$ being the odd extension of $\varphi(x)$ onto $\mathbb{R}$ (see Fig. 5.8):

$$\hat{u}(0,x) = \hat{\varphi}(x) \equiv \begin{cases} \varphi(x), & x \geq 0; \\ -\varphi(x), & x < 0. \end{cases}$$

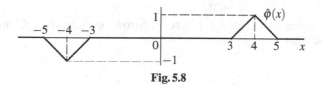

Fig. 5.8

Set $u(x,t) \equiv \hat{u}(x,t)\big|_{x \geq 0}$. Obviously, u satisfies equation (5.1) and the initial data (5.2). Below, we will see that the boundary condition (5.3) is also satisfied, since $\hat{u}(x,t)$ is odd in x. The region $x < 0$ is virtual, or nonphysical.

Construction of $\hat{u}(x,t)$. According to the d'Alembert formula (2.20),

$$\hat{u}(x,t) = \frac{\hat{\varphi}(x-t)}{2} + \frac{\hat{\varphi}(x+t)}{2},$$

that is, we need to divide $\hat{\varphi}(x)$ by two, shift by t to the right and to the left, and to add up the results. See Fig. 5.9.

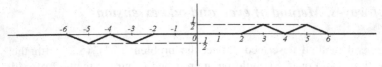

The string at $t = 1$.

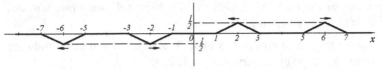

$t = 2$. Arrows indicate the direction of the motion of the humps.

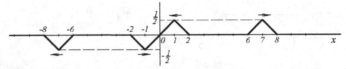

$t = 3$. The left hump in the physical region $x > 0$ approaches the nail at $x = 0$.

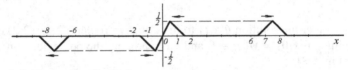

$t = 3.5$. The nail pulls the hump over.

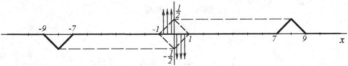

$t = 4$. The deviation for $-1 \leq x \leq 1$ is zero. Arrows indicate the velocities of the points of the string.

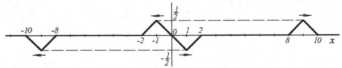

$t = 5$. The humps have parted (the arrows indicate the directions of motion of humps).

Fig. 5.9

And so on: in the physical region $x > 0$ the two humps move to the right (while in the virtual region $x < 0$ the two humps move to the left).

Remark 5.14. The boundary condition (5.3) at $x = 0$ holds for all $t > 0$ since $\hat{u}(x,t)$ is odd in x.

Let us consider oscillations of a piano string after having been hit with a hammer.

Problem 5.15. Solve the problem (5.1)–(5.3) with $a = 1$ and the initial data as on Fig. 5.10. Plot the string at $t = 1, 2, 3, 4, 5$, and 6.

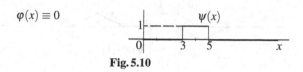

Fig. 5.10

Solution. Let us set $\hat{\varphi}(x) \equiv 0$, $x \in \mathbb{R}$, and let us extend $\psi(x)$ onto $\mathbb{R}$ so that it is odd:

$$\hat{\psi}(x) = \begin{cases} \psi(x), & x \geq 0; \\ -\psi(-x), & x < 0. \end{cases}$$

This function is plotted on Fig. 5.11.

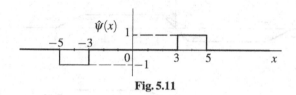

Fig. 5.11

Consider the solution $\hat{u}$ to the Cauchy problem (2.1)–(2.2) with the initial data $\hat{\varphi}$ and $\hat{\psi}$. By the d'Alembert formula (2.20),

$$\hat{u}(x,t) = \hat{\phi}(x+t) - \hat{\phi}(x-t), \tag{5.22}$$

where the function $\hat{\phi}(x) \equiv \frac{1}{2} \int_{-\infty}^{x} \hat{\psi}(y)\, dy$ is plotted on Fig. 5.12.

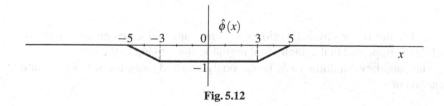

Fig. 5.12

We set $u(x,t) \equiv \hat{u}(x,t)\big|_{x>0}$. Obviously, $u(x,t)$ satisfies (5.1) and (5.2). As will be seen below, the boundary condition (5.3) is also satisfied. Construction of $\hat{u}(x,t)$ according to formulas (5.22) is on Fig. 5.13.

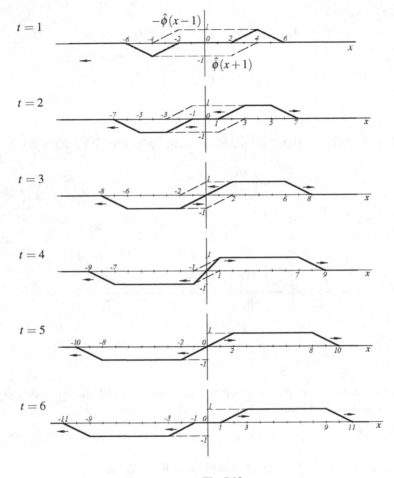

Fig. 5.13

And so on: In the physical region $x > 0$ the trapezoid keeps moving to the right (while the trapezoid in the unphysical region keeps moving to the left).

The boundary condition (5.3) is obviously satisfied since the solution is an odd function of x.

Problem 5.16. Draw the shape of the string at $t = 3.5$ and $t = 4.5$.

Let us now consider *the method of even extension*.

Problem 5.17. Solve the mixed problem (5.1), (5.2), (5.8) with $a = 1$ and the initial data as on Fig. 5.10. Draw the shape of the string at $t = 1, 2, 3, 3.5, 4, 4.5, 6$.

Hint. Use the even extension for $\varphi(x)$ and $\psi(x)$. Then $\hat{u}(x,t)$ will be even in x, hence the boundary condition (5.8) will be satisfied.

Problem 5.18. For $t < 0$, there is a wave of deformation propagating to the left along an elastic semi-infinite rod:

$$u(x,t) = \begin{cases} \sin(x+3t), & x > -3t; \\ 0, & 0 < x < -3t, \quad t < 0. \end{cases} \tag{5.23}$$

The left end of the rod at $x = 0$ is elastically attached (see (1.18)):

$$0 = -2u(0,t) + 3\frac{\partial u}{\partial x}(0,t), \quad t > 0. \tag{5.24}$$

Find $u(x,t)$ for $t > 0$.

Solution. As it follows from the condition of the problem,

$$\frac{\partial^2 u}{\partial t^2} = 9\frac{\partial^2 u}{\partial x^2}, \quad x > 0, \, t > 0; \qquad u(x,0) = \sin x, \quad \frac{\partial u}{\partial t}(x,0) = 3\cos x, \quad x > 0.$$

From here, for $x > 3t$, the d'Alembert formula yields

$$u(x,t) = \sin(x+3t), \qquad x > 3t, \tag{5.25}$$

as in (5.23). In the region $x < 3t$ we are looking for a solution in the form $u(x,t) = f(x-3t) + \sin(x+3t)$. Substituting this into the boundary condition (5.24), we get

$$0 = -2f(-3t) - 2\sin 3t + 3f'(-3t) + 3\cos 3t.$$

The substitution $y = -3t$ gives

$$3f'(y) - 2f(y) = -2\sin y - 3\cos y, \qquad y < 0. \tag{5.26}$$

Therefore, for $y < 0$, $f(y) = Ce^{2y/3} + A\cos y + B\sin y$. The constants A and B are found by substituting $f(y)$ into (5.26):

$$-3A\sin y + 3B\cos y - 2A\cos y - 2B\sin y = -2\sin y - 3\cos y.$$

Thus, $-3A - 2B = -2$, $3B - 2A = -3$, leading to $A = 12/13$ and $B = -5/13$. The value of C is found from the continuity condition (5.12) at the characteristic $y = x - 3t = 0$:

$$C \mid A = f(0-) - f(0+) = 0, \qquad C = -12/13.$$

Answer. $u(x,t) = \begin{cases} \sin(x+3t), & x > 3t; \\ \dfrac{-12e^{\frac{2}{3}(x-3t)} + 12\cos(x-3t) - 5\sin(x-3t)}{13} + \sin(x+3t), & x < 3t. \end{cases}$

6 Finite string

The d'Alembert method

Transversal oscillations of a finite string in the absence of external forces are described by the equation

$$\frac{\partial^2 u(x,t)}{\partial t^2} = a^2 \frac{\partial^2 u}{\partial x^2}, \qquad 0 < x < l, \quad t > 0. \tag{6.1}$$

For the determination of the motion of the string we need the initial data

$$u(x,0) = \varphi(x), \qquad \dot{u}(x,0) = \psi(x), \qquad 0 < x < l \tag{6.2}$$

and the boundary conditions at the ends. For example, if the ends are fixed, then

$$u(0,t) = 0 \quad \text{and} \quad u(l,t) = 0, \qquad t > 0. \tag{6.3}$$

Solution of the mixed problem (6.1)–(6.3) could be found by the d'Alembert method along the lines of Section 5, as follows.

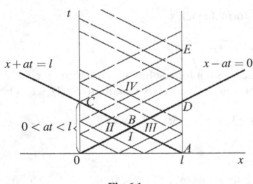

Fig. 6.1

A. Substituting (5.4) into the initial data (6.2) at $t = 0$, $0 < x < l$, we find by formulas (2.18), (2.19) the waves $f(x-at)$ and $g(x+at)$ at these points. This gives the solution $u(x,t)$ in region I (the triangle OAB) on Fig. 6.1.

B. Substituting the ansatz (5.4) into the boundary condition (6.3) at $x = 0$, we find the reflected wave $f(x-at)$ from knowing the incident wave $g(x+at)$ at the points of the interval OC. This gives the solution $u(x,t)$ in region II (the triangle OBC) on Fig. 6.1.

C. Substituting (5.4) into (6.3) for $x = l$, we find the reflected wave $g(x+at)$ from knowing the incident wave $f(x-at)$ at the points of the interval AE.

And so on. This allows us to find the solution $u(x,t)$ in the entire semi-strip $0 < x < l$, $t > 0$, successively decomposing it into regions, bounded by characteristics similar to characteristics OD, AC, and CE. In the same fashion one can solve the mixed problem (6.1)–(6.2) with boundary conditions other than (6.3).

Remark 6.1. The asymptotic properties of solutions to the problem (6.1)–(6.3) as $t \to \infty$, and, in particular, the frequencies of oscillations, are easier to investigate using the Fourier method, which is described in Chapter 2.

Method of even and odd extension

Problem 6.2. Solve the problem (6.1)–(6.3) for $a = 1, l = 6$ and the initial data $u(x,0) = \varphi(x)$ from Fig. 6.2, $\dot{u}(x,0) = \psi(x) = 0$. Plot the shape of the string for $t = 1, 2, \ldots$ and find the period T of the oscillations.

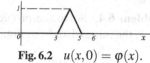

Fig. 6.2 $u(x,0) = \varphi(x)$.

Solution. See Fig. 6.3.

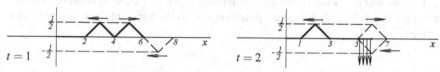

We send a virtual hump from the right.

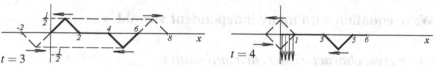

We send a virtual hump from the left.

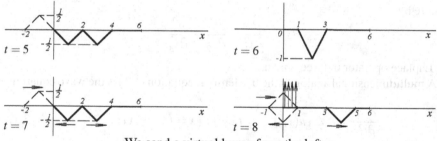

We send a virtual hump from the left.

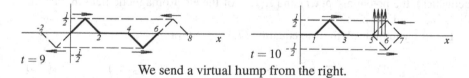

We send a virtual hump from the right.

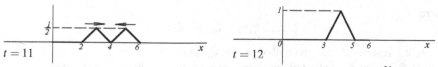

We see that the process is periodic, with the period $T = 12 = \frac{2l}{a}$.

Fig. 6.3

Problem 6.3 (The piano string). Solve the problem (6.1)–(6.3) for the string of length $l = 6$, with $a = 1$, and the initial data from Fig. 5.10. Plot the shape of the string at $t = 1, 2, \ldots$ and find the period of oscillations.

Problem 6.4. Solve the problem (6.1)–(6.2) for the string of length $l = 6$, with $a = 1$, the initial data from Fig. 5.10, and with the boundary conditions

$$\frac{\partial u}{\partial x}(0,t) = 0 \quad \text{and} \quad \frac{\partial u}{\partial x}(l,t) = 0, \qquad t > 0.$$

Plot the shape of the string at $t = 1, 2, \ldots$ and find the period of oscillations.

Hint. One should apply the method of even reflections, that is, send reflected virtual humps (see Fig. 6.3) with the same "polarization" as the incident ones (not with the opposite).

7 Wave equation with many independent variables

Plane waves, characteristics, discontinuities

Denote by

$$\triangle_3 \equiv \frac{\partial^2}{\partial x_1^2} + \frac{\partial^2}{\partial x_2^2} + \frac{\partial^2}{\partial x_3^2}$$

the Laplace operator in three dimensions.

A multidimensional analog of the d'Alembert equation (1.1) is the wave equation

$$\frac{\partial^2 u}{\partial t^2} = a^2 \triangle_3 u(x,t), \qquad t > 0, \quad x = (x_1, x_2, x_3) \in \mathbb{R}^3, \qquad (7.1)$$

where $a > 0$. This equation describes the air pressure $p(x,t)$ (the sound wave in acoustics), the potentials $\varphi(x,t)$ and $A(x,t)$ of the electromagnetic field in electrodynamics, and so on.

Let us try to find solutions to equation (7.1) in the form

$$u(x,t) = f(\tau t + \xi_1 x_1 + \xi_2 x_2 + \xi_3 x_3) = f\left(\tau t + \langle \xi, x \rangle\right), \qquad (7.2)$$

where

$$\langle \xi, x \rangle \equiv \xi_1 x_1 + \xi_2 x_2 + \xi_3 x_3.$$

We are interested in solutions with $\xi = (\xi_1, \xi_2, \xi_3) \neq 0$.

Such functions are called plane waves. Here is the reason for this name:

a. At fixed $t = t_0$, the level surfaces $u(x,t_0) = \text{const}$ are represented by the planes

$$\tau t + \langle \xi, x \rangle = c$$

orthogonal to the vector $\eta = (\tau, \xi)$;

b. For $t = t_0$ and $t = t_1$, with $t_0 \neq t_1$, the function $u(x, t_1)$ differs from $u(x, t_0)$ by the shift represented by the vector

$$-\frac{\xi}{|\xi|^2}\tau(t_1 - t_0).$$

Indeed,

$$u\left(x + \frac{\xi}{|\xi|^2}\tau(t_1 - t_0), t_0\right) = f\left(\tau t_0 + \left\langle \xi, x + \frac{\xi}{|\xi|^2}\tau(t_1 - t_0)\right\rangle\right)$$

$$= f\left(\tau t_0 + \langle \xi, x \rangle + \frac{\langle \xi, \xi \rangle}{|\xi|^2}\tau(t_1 - t_0)\right) = f(\tau t_1 + \langle \xi, x \rangle) = u(x, t_1).$$

Thus, (7.2) is a wave moving along the direction of the vector $-\xi$ with the speed

$$v = \frac{\tau}{|\xi|}.$$

We denote the unit vector in the direction $-\xi$ by

$$\omega = -\frac{\xi}{|\xi|}.$$

Then $\tau = v|\xi|$, $\xi = -\omega|\xi|$, and, therefore, (7.2) can be written as

$$u(x, t) = f\left(v|\xi|t - \langle \omega, x \rangle |\xi|\right) = f\left((vt - \langle \omega, x \rangle)|\xi|\right) = g(vt - \langle \omega, x \rangle),$$

where

$$|\omega| = 1, \qquad g(\lambda) \equiv f(\lambda|\xi|), \qquad \lambda \in \mathbb{R}.$$

After these preliminary remarks let us proceed to finding the solution to equation (7.1) in the form (7.2). We substitute (7.2) into (7.1), and, using the chain rule, we get:

$$f''(\tau t + \langle \xi, x \rangle)\tau^2 = a^2 f''(\tau t + \langle \xi, x \rangle)(\xi_1^2 + \xi_2^2 + \xi_3^2). \tag{7.3}$$

Assuming that $f''(z) \not\equiv 0$, we get from (7.3) the characteristic equation

$$\tau^2 = a^2|\xi|^2. \tag{7.4}$$

Solutions of this equation are vectors $\eta = (\tau, \xi) \in \mathbb{R}^4$, lying on the (three-dimensional) cone Q in $\mathbb{R}^4$, whose base is a two-dimensional sphere

$$\tau = 1, \qquad |\xi| = \frac{1}{a}.$$

See Fig. 7.1.

Conversely, for any $\eta = (\tau, \xi) \in \mathbb{R}^4$ satisfying (7.4), the plane wave (7.2) with any function $f(z)$ is a solution to equation (7.1).

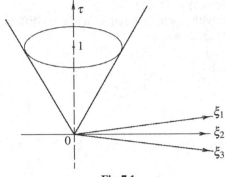

Fig. 7.1

In particular, $f(z)$ could be taken discontinuous (or rapidly changing) at some point, for example, at $z = 2$ (see Fig. 3.9). Then the solution (7.2) will have the same discontinuity (or rapid change) along the entire hyperplane in $\mathbb{R}^4_{x,t}$ (if $\eta = (\tau, \xi) \neq 0$):

$$\tau t + \langle \xi, x \rangle = 2. \tag{7.5}$$

For fixed t this discontinuity is located on the plane in $\mathbb{R}^3_x$ described by equation (7.5). As the time increases, this plane moves in the direction of its normal, represented by $-\xi$, with the speed $v = \frac{|\tau|}{|\xi|} = a$ (see (7.4)).

Definition 7.1 (Characteristic conormals, hyperplanes, and hypersurfaces).

a. *A vector*

$$\eta = (\tau, \xi_1, \xi_2, \xi_3) \in \mathbb{R}^4, \qquad \eta \neq 0,$$

satisfying (7.4) is called a characteristic conormal of the wave equation (7.1).

b. *A hyperplane*

$$\eta^\perp = \left\{ (t, x) \in \mathbb{R}^4 : \tau t + \langle \xi, x \rangle = \text{const} \right\},$$

orthogonal to a particular characteristic conormal $\eta = (\tau, \xi)$, is called a characteristic hyperplane of the wave equation (7.1).

c. *A hypersurface in $\mathbb{R}^4$ is called a characteristic hypersurface if the tangent hyperplane at each point is characteristic.*

Remark 7.2. Due to the characteristic equation (7.4), the speed of propagation of all the plane waves which satisfy the wave equation (7.1) is equal to a:

$$v^2 = \frac{\tau^2}{|\xi|^2} = a^2. \tag{7.6}$$

Conclusion. Any characteristic hyperplane could be a surface of discontinuity of solutions to equation (7.1). See Remark 2.1.

All the plane waves satisfying equation (7.1) propagate with the speed a.

It is the formula (7.6) that the discovery of *the electromagnetic nature of light* and *the special theory of relativity* is connected with. From the equations of electro-dynamics, Maxwell derived that the potentials of the electromagnetic field satisfy the wave equation (7.1) with the coefficient

$$a^2 = \frac{1}{\varepsilon_0 \mu_0}. \tag{7.7}$$

Here ε_0 and μ_0 are the electric and magnetic permeability of vacuum, respectively, which are found experimentally from purely electromagnetic measurements. When Maxwell computed the speed of propagation of the electromagnetic waves, it turned out that this speed with the great accuracy coincided with the speed of light:

$$a = \frac{1}{\sqrt{\varepsilon_0 \mu_0}} \approx 299\,792\,\frac{km}{s}.$$

This led Maxwell to the conclusion that the light also has an electromagnetic nature!

The special theory of relativity was another great discovery related to formulas (7.6) and (7.7). The question naturally arises: In what reference frame is the value of the speed of light actually equal to $\frac{1}{\sqrt{\varepsilon_0 \mu_0}}$? It is known that all the laws of mechanics are the same in any inertial reference frame. Thus, it is natural to assume that the laws of electrodynamics also hold in any inertial reference frame. But, according to (7.7), the speed of light should also be the same in all such systems! Such a property of the velocity, though, contradicts Newton's mechanics. It follows that either the Maxwell equations are only valid in a particular reference frame, related to the sta-tionary "ether," or the Newton laws of mechanics are not exact. It is for settling this question that Michelson and Morley built their famous experiment to justify that the speed of light is the same in different inertial reference frames, and, consequently, to prove the absence of the stationary "ether" and inexactness of Newton's mechanics (at high speeds). The necessary refinement of the laws of mechanics was later given by Einstein.

The domain of dependence. The Kirchhoff formula

Let us try to find the domain of dependence for equation (7.1) with the aid of cha-racteristics, as in Section 4 (Fig. 4.1). That is, let us consider the Cauchy problem for equation (7.1) with initial conditions at $t = 0$:

$$u\big|_{t=0} = \varphi(x), \qquad \frac{\partial u}{\partial t}\bigg|_{t=0} = \psi(x), \qquad x \in \mathbb{R}^3. \tag{7.8}$$

Let us draw all the characteristics (*the characteristic hyperplanes*) of equation (7.1) through a particular point $(x_0, t_0) \in \mathbb{R}^4$, $t_0 > 0$. On Fig. 7.2, η is a characteristic

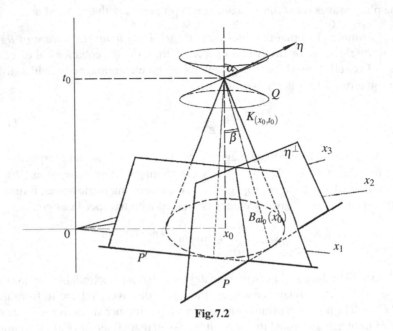

Fig. 7.2

conormal, $\eta^\perp$ is the characteristic hyperplane orthogonal to η and passing through (x_0, t_0). This characteristic hyperplane intersects the "initial" hyperplane $t = 0$ along the plane P (represented by a line on Fig. 7.2).

Conjecture 7.3. The domain of dependence for a solution at (x_0, t_0) is the part of the hyperplane $t = 0$ bounded by the planes $P, P', \ldots$ (it is analogous to Fig. 4.1).

This region is a ball of radius at_0 centered at x_0, denoted by $B_{at_0}(x_0)$. To see this, one should notice the following: The normal η belongs to the characteristic cone Q described by equation (7.4), while the hyperplane $\eta^\perp$ is tangent to the cone $K_{(x_0, t_0)}$ which is dual to the characteristic cone Q (see Fig. 7.2). Therefore the planes P, P', ... are tangent to the base of the cone $K_{(x_0, t_0)}$, which is represented by the sphere $S_{at_0}(x_0) = \{x \in \mathbb{R}^3 : |x - x_0| = at_0\}$.

Remark 7.4. The cone $K_{(x_0, t_0)} = \{(x, t) \in \mathbb{R}^4 : |x - x_0| = a|t - t_0|\}$ is called the *light cone* of equation (7.1) at the point (x_0, t_0). Its *dual cone* $Q = \{(\tau, \xi) \in \mathbb{R} \times \mathbb{R}^3 : |\tau| = a|\xi|\}$ is called the *characteristic cone* of equation (7.1) at the point (x_0, t_0). The cones $K_{(x_0, t_0)}$ and Q are orthogonal, that is, $\alpha + \beta = \frac{\pi}{2}$. See Fig. 7.2.

The intersection of the cone $K_{(x_0, t_0)}$ with the hyperplane $t = 0$ is a sphere given by

$$S_{at_0}(x_0) = \{x \in \mathbb{R}^3 : |x - x_0| = at_0\}.$$

Thus, Conjecture 7.3 means that the domain of dependence for u at the point (x_0, t_0) is a ball of radius at_0 centered at x_0, $B_{at_0}(x_0) = \{x \in \mathbb{R}^3 : |x - x_0| = at_0\}$. This conjecture is equivalent to saying that all the solutions of equation (7.1) propagate with the speed a. Let us point out that we already proved this for the plane waves.

Our conjecture is correct indeed. Moreover, it turns out that the domain of dependence is smaller than a ball: It only consists of the sphere $S_{at_0}(x_0)$. In particular, this follows from the Kirchhoff formula for the solution to the Cauchy problem (7.1), (7.8):

$$u(x,t) = \frac{1}{4\pi a^2 t} \int\limits_{|y-x|=at} \psi(y)\,dS_y + \frac{\partial}{\partial t}\left(\frac{1}{4\pi a^2 t} \int\limits_{|y-x|=at} \varphi(y)\,dS_y\right). \qquad (7.9)$$

For the derivation of this formula, see [Pet91].

Propagation of waves. The Huygens principle

Problem 7.5. The function $u(x,t)$ solves (7.1) with $a = 1$ and with the initial data such that

$$\varphi(x) \equiv \psi(x) \equiv 0 \quad \text{for} \quad |x| > 1.$$

Find where (for certain) $u(x,t) \equiv 0$ at $t = 1, 2, 3, 4$.

Solution. First, assume that a is arbitrary. Then $u(x,t) = 0$ if the region of integration in (7.9), that is, the sphere $|y - x| = at$, does not intersect the region $|y| \leq 1$ where $\varphi(y)$ and $\psi(y)$ are supported. See Fig. 7.3. Clearly, this condition is equivalent to

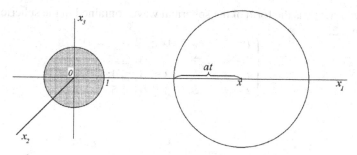

Fig. 7.3

one of the following two conditions: either, as on Fig. 7.3, the ball $|y| \leq 1$ is outside the sphere $|x - y| = at$,

$$1 + at < |x|, \qquad (7.10)$$

or, as on Fig. 7.4, the sphere $|y - x| = at$ contains the ball $|y| \leq 1$ strictly inside:

$$at > 1 + |x|. \qquad (7.11)$$

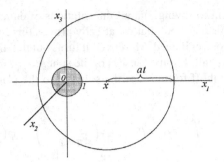

Fig. 7.4

The condition (7.10) with $a = 1$ yields the identity $u(x,t) \equiv 0$ in the following regions:

$$\begin{cases} t = 1 & \Rightarrow & |x| > 2; \\ t = 2 & \Rightarrow & |x| > 3; \\ t = 3 & \Rightarrow & |x| > 4; \\ t = 4 & \Rightarrow & |x| > 5. \end{cases}$$

The condition (7.11) with $a = 1$ yields the identity $u(x,t) \equiv 0$ in the following regions:

$$\begin{cases} t = 1 & \Rightarrow & x \in \emptyset; \\ t = 2 & \Rightarrow & |x| < 1; \\ t = 3 & \Rightarrow & |x| < 2; \\ t = 4 & \Rightarrow & |x| < 3. \end{cases}$$

Therefore, $u(x,t)$ has the form of the spherical wave contained in the spherical layer of thickness 2:

$$\begin{cases} t = 1 & \Rightarrow & |x| \leq 2; \\ t = 2 & \Rightarrow & 1 \leq |x| \leq 3; \\ t = 3 & \Rightarrow & 2 \leq |x| \leq 4; \\ t = 4 & \Rightarrow & 3 \leq |x| < 5. \end{cases} \tag{7.12}$$

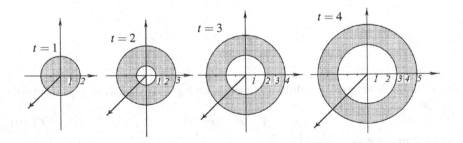

Fig. 7.5

Answer. $u(x,t)$ is for certain equal to zero outside the spherical layers (7.12) (although it could also be equal to zero somewhere inside these layers). See Fig. 7.5.

Conclusion. As seen from (7.12), the front of the spherical wave propagates with the speed 1. In the case of an arbitrary a, it is seen from (7.10) and (7.11) that the solution $u(x,t)$ could only be different from zero in a spherical layer

$$at - 1 \leq |x| \leq at + 1$$

of thickness 2. This wave has two fronts: the forward front $|x| = at + 1$ and the rear front $|x| = at - 1$, both propagating with the speed a.

Problem 7.6. Given: $a = 1$, $\varphi(x) \equiv \psi(x) \equiv 0$ at $|x| < 2$ or $|x| > 4$ (as on Fig. 7.5 for $t = 3$). Where is $u(x,t)$ identically equal to zero for $t = 1, 2, 3, 4, 5$?

Solution. There are three possibilities, I, II, and III (see Fig. 7.6), of the location of the sphere $|y - x| = t$ so that we would have $u(x,t) \equiv 0$. For location I, analogously

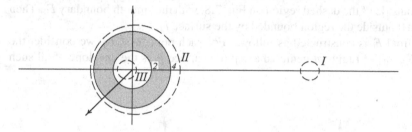

Fig. 7.6

to (7.10), in the case of a general value of a, $4 + at < |x|$. For location II, analogously to (7.11), $at > 4 + |x|$. Finally, for location III, $|x| + at < 2$.

Since we are given $a = 1$, we get the following:

a. At $t = 1$ the sphere $|y - x| = t$ is of radius 1 and locations I and III are possible, while II is not. As a result, we see that $u(x, 1)$ is supported somewhere inside the layer $1 \leq |x| \leq 5$. See Fig. 7.7. This result seriously differs from Fig. 7.5 at $t = 4$.
b. At $t = 2$ the radius of the sphere of integration is equal to 2, therefore, only location I is possible. Therefore, the solution is supported somewhere inside the ball $|x| \leq 6$.
c. At $t = 3$ it is also only location I which is possible (the sphere of integration is of radius 3), therefore the solution is supported inside the ball $|x| \leq 7$.
d. The same happens for $t = 4$: $u(x, 4)$ may be different from zero only inside the ball $|x| \leq 8$.

e. Finally, at $t = 5$, in addition to location I, location II also becomes possible (the sphere of integration is of radius 5), etc. See Fig. 7.7.

We now see that $u(x,t)$ for $t > 4$ is supported inside an expanding spherical wave of thickness 8.

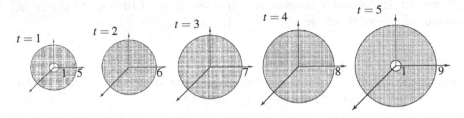

Fig. 7.7

The *Huygens principle* is the rule which allows us to build the forward front F_t of the wave at the moment t if it is known at $t = 0$. This rule follows from the Kirchhoff formula (7.9) and consists of the following: Let $u\big|_{t=0}$ and $\dot{u}\big|_{t=0}$ be equal to zero outside of the dashed region on Fig. 7.8, with the smooth boundary F_0. Then $u(x,t) \equiv 0$ outside the region bounded by the surface F_t.

The front F_t is constructed as follows: For each point $x_0 \in F_0$, we consider the sphere $S_{at}(x_0)$ of radius at centered at x_0; the surface F_t is the envelope of all such spheres.

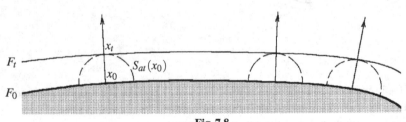

Fig. 7.8

Let us assume that there is a unique point where the front F_t touches the sphere $S_{at}(x_0)$, and denote this point by x_t. It is easy to see that the interval $[x_0, x_t]$ is orthogonal to F_t if F_t is a smooth surface. One can also check that $[x_0, x_t] \perp F_0$ (check this!). Consequently, the front F_t could also be constructed in the following way: From each point $x_0 \in F_0$ we draw an interval $[x_0, x_t] \perp F_0$ of length at. The front F_t is then the set of all such points x_t. The intervals $[x_0, x_t]$ are called *the light rays*. Therefore, the Huygens principle implies that

‖ *The waves propagate along the light rays.*

Diffusion of waves in two dimensions. The Poisson formula

The wave equation in the plane,

$$\frac{\partial^2 u}{\partial t^2}(x,t) = a^2 \triangle_2 u \equiv a^2 \left(\frac{\partial^2 u}{\partial x_1^2} + \frac{\partial^2 u}{\partial x_1^2}\right), \qquad x \in \mathbb{R}^2, \quad t > 0, \tag{7.13}$$

is obtained from (7.1) when $u(x_1,x_2,x_3,t)$ does not depend on x_3. This is the case when neither the initial data nor the external sources (such as the current or charges in electrodynamics or the sound sources in acoustics) depend on x_3. For example, the potentials of the magnetic field generated by the current in a straight wire and acoustic field of a long straight highway satisfy equation (7.13). The waves $u(x_1,x_2,t)$, which do not depend on x_3, are called cylindrical.

In this case, the initial data φ and ψ also do not depend on x_3:

$$u\big|_{t=0} = \varphi(x), \qquad \dot{u}\big|_{t=0} = \psi(x), \qquad x \in \mathbb{R}^2. \tag{7.14}$$

Solution to the problem (7.13)–(7.14) is given by the Poisson formula

$$u(x,t) = \frac{1}{2\pi a} \int\limits_{|y-x|<at} \frac{\psi(y)\,dy}{\sqrt{(at)^2 - |y-x|^2}} + \frac{1}{2\pi a}\frac{\partial}{\partial t} \int\limits_{|y-x|<at} \frac{\varphi(y)\,dy}{\sqrt{(at)^2 - |y-x|^2}}. \tag{7.15}$$

These integrals are evaluated over the disc $|y-x| < at$ and not over its boundary; this is different from the Kirchhoff formula (7.9). Consequently, the propagation of the cylindrical waves (or simply "the plane waves") is different from that of the spherical waves.

Problem 7.7. The solution $u(x,t)$ to (7.13) with $a = 1$ satisfies the initial data (7.14) with $\varphi(x) \equiv \psi(x) \equiv 0$ at $|x| > 1, x \in \mathbb{R}^2$. Where does one have $u(x,t) \equiv 0$ for $t = 1,2,3,4,5$?

Answer. $|x| > 2$ for $t = 1$; $|x| > 3$ for $t = 2$; $|x| > 4$ for $t = 3$; $|x| > 5$ for $t = 4$.

Remark 7.8. In this problem the cylindrical wave has the forward front but does not have the rear front, contrary to the spherical waves in two previous problems. This phenomenon is called the diffusion of waves. It turns out that for all odd $n \geq 3$ the wave equation with n spatial variables $x_1, \ldots, x_n$ has both the forward and rear fronts, while for all even $n \geq 2$ (and for $n = 1$ as well!) there is the forward front but no rear front.

Remark 7.9. If in the last problem the functions φ and ψ which enter (7.14) are bounded, then the solution converges to zero: $u(x,t) \to 0$ for $t \to \infty$, $\forall x \in \mathbb{R}^2$. This is seen from (7.15). (Prove this!)

Remark 7.10 ("The method of descent" from $n = 3$ to $n = 2$). One can obtain the Poisson formula (7.15) from the Kirchhoff formula (7.9) using the independence of φ and ψ on x_3 (see [Pet91]).

8 General hyperbolic equations

General hyperbolic equations with constant coefficients

Let us first consider the equation $Au = 0$ where A is a *homogeneous differential operator*, such that all the terms are the partial derivatives of the same total order m:

$$Au(y) \equiv \sum_{|\alpha|=m} a_\alpha \partial_y^\alpha u(y) = 0, \qquad y = (y_1, \ldots, y_n) \in \mathbb{R}^n. \tag{8.1}$$

Above, $\alpha = (\alpha_1, \ldots, \alpha_n)$, $\alpha_k = 0, 1, 2, \ldots$; $|\alpha| \equiv \alpha_1 + \cdots + \alpha_n$; $\partial_y^\alpha \equiv \frac{\partial^{|\alpha|}}{\partial y_1^{\alpha_1} \ldots \partial y_n^{\alpha_n}}$.
Let us look for the solutions of the type of the plane waves:

$$u(y) = f(\langle \eta, y \rangle) = f(\eta_1 y_1 + \ldots + \eta_n y_n), \qquad y \in \mathbb{R}^n, \tag{8.2}$$

where f is a function of one variable. Substituting (8.2) into (8.1) we get:

$$\sum_{|\alpha|=m} a_\alpha \eta^\alpha f^{(m)}(\langle \eta, y \rangle) = 0, \qquad \text{where} \quad \eta^\alpha \equiv \eta_1^{\alpha_1} \ldots \eta_n^{\alpha_n}. \tag{8.3}$$

From here, assuming that $f^{(m)}(z) \not\equiv 0$, we get, analogously to (7.4), the algebraic equation of the characteristics (compare with (4.28)):

$$\tilde{A}(\eta) \equiv \sum_{|\alpha|=m} a_\alpha \eta^\alpha = 0. \tag{8.4}$$

This equation defines the cone Q in $\mathbb{R}^n$, that is, if $\eta \in Q$, then also $\lambda \eta \in Q$ for all $\lambda \in \mathbb{R}$. Q is called the *characteristic cone*.

We see from (8.3) that the plane wave (8.2) with an arbitrary function f satisfies the differential equation (8.1) if and only if η satisfies (8.4).

Definition 8.1 (Characteristic conormals, hyperplanes, and hypersurfaces).

a. *A vector $\eta \in \mathbb{R}^n$, $\eta \neq 0$ satisfying (8.4) is called a characteristic conormal of the differential equation (8.1);*
b. *A hyperplanes $\eta^\perp \equiv \{y \in \mathbb{R}^n : \langle \eta, y \rangle = \text{const}\}$ orthogonal to a characteristic conormal η is a characteristic hyperplane of the differential equation (8.1);*
c. *A hypersurface in $\mathbb{R}^n$ is called a characteristic hypersurface of equation (8.1) if the tangent hyperplanes at all its points are characteristic.*

Definition 8.2. *Equation (8.1) is called (strictly) hyperbolic in the direction of the axis Oy_1 if equation (8.4) on η_1 for any fixed $\eta' \equiv (\eta_2, \ldots, \eta_n) \in \mathbb{R}^{n-1} \setminus 0$ has exactly m different real roots $\eta_1^{(k)} = \lambda_k(\eta'), k = 1, \ldots, m$:*

$$\lambda_1(\eta') < \ldots < \lambda_m(\eta'). \tag{8.5}$$

Geometrically, (8.5) means that the cone Q consists of exactly m different sheets.

Example 8.3. For the wave equation (7.1) its order is $m = 2$ and dimension is $n = 4$. Equation (8.4) with the variables $(\eta_1, \eta_2, \eta_3, \eta_4)$ can be written in the form (7.4) with the variables $(\tau, \xi_1, \xi_2, \xi_3)$. Equation (7.4) has two roots $\tau = \pm a|\xi|$; hence,

$$\lambda_1 = -a|\xi|, \qquad \lambda_2 = a|\xi|; \qquad \xi \in \mathbb{R}^3 \setminus 0.$$

The cone Q consists of two sheets; see Fig. 8.1. Thus, the wave equation is hyperbolic in the variable t.

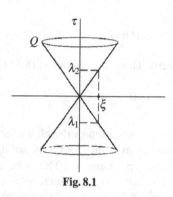

Fig. 8.1

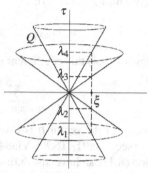

Fig. 8.2

Example 8.4. Let us consider the equation of order $m = 4$:

$$\left(\frac{\partial^2}{\partial t^2} - \Delta\right)\left(\frac{\partial^2}{\partial t^2} - 9\Delta\right)u(x,t) = 0, \qquad x \in \mathbb{R}^3, \quad t > 0.$$

The characteristic equation (8.4) takes the form

$$(\tau^2 - |\xi|^2)(\tau^2 - 9|\xi|^2) = 0.$$

It has 4 roots: $\tau = \pm|\xi|$ and $\tau = \pm 3|\xi|$, and hence

$$\lambda_1 = -3|\xi|, \quad \lambda_2 = -|\xi|, \quad \lambda_3 = |\xi|, \quad \lambda_4 = 3|\xi|; \qquad \xi \in \mathbb{R}^3 \setminus 0.$$

Therefore, the cone Q consists of four sheets (see Fig. 8.2).

Question 8.5. Is the strict hyperbolicity condition related to the condition (4.8)?

Answer. For the second order equations with two independent variables they are equivalent. Indeed, for the equation (4.6), the characteristic equation (8.4) has the form

$$A(\tau, \xi) \equiv a\tau^2 + 2b\tau\xi + c\xi^2 = 0.$$

Under the condition (4.8), its roots $\tau = \frac{b \pm \sqrt{D}}{a}\xi$ are real and different.

Taking in (8.2) a discontinuous function $f(z)$, we see that the solution to equation (8.1) can have a discontinuity along any given characteristic hyperplane (see Remark 2.1).

Remark 8.6. Let us take the direction of the characteristic conormal η as a new coordinate axis, so that the plane $y_1 = 0$ coincides with $\eta^\perp$, while other coordinate axes $y_2, \ldots, y_n$ are chosen arbitrarily, as long as they correspond to a linear nondegenerate change of variables. Then, as turns out (prove this!), equation (8.1) in the new coordinates contains the term $b_{(m,0,\ldots,0)} \frac{\partial^m u}{\partial y_1^m}$ with the following coefficient (compare with (4.24)–(4.27)):

$$b_{(m,0,\ldots,0)} = \tilde{A}(\mathrm{grad}\, y_1) = C\tilde{A}(\eta).$$

But in view of (8.4) this coefficient is equal to zero. Therefore, equation (8.1) takes the form

$$\sum_{|\alpha|=m, \alpha_1 \leq m-1} b_\alpha \partial_y^\alpha u(y) = 0. \tag{8.6}$$

This property of the vector η is usually taken as the definition of the characteristic conormal (see [Pet91, TS90, Vla84]). It is transparent from (8.6) why solutions to equation (8.1) can have discontinuities along the hyperplane $\eta^\perp$. This is because each term in equation (8.6) contains at least one derivative with respect to $y_2, \ldots, y_n$. Consequently, any function of y_1 satisfies equations (8.6) and (8.1); in particular, it could be a discontinuous function of y_1 (compare with Remark 4.1).

Now let us consider the equation $Au = 0$ where A is a general *nonhomogeneous* operator:

$$\sum_{|\alpha|\leq m} a_\alpha \partial_y^\alpha u(y) = 0, \qquad y \in \mathbb{R}^n. \tag{8.7}$$

We no longer know the solutions to this equation in the form of the plane waves. But, by the definition, it is accepted that the characteristic equation for (8.7) is (8.4), that is, we omit the lower order terms.

We know that solutions to equation (8.1) could have discontinuities along any given characteristic hyperplane. It turns out that this is also the case for equation (8.7) if it is strictly hyperbolic. Example 8.8 below shows that if the hyperbolicity condition is not satisfied, this may no longer be the case!

Examples of nonhyperbolic equations

The heat equation reads:

$$\frac{\partial u}{\partial t} = a^2 \triangle u(x,t), \qquad x \in \mathbb{R}^3, \quad t > 0. \tag{8.8}$$

For this equation the characteristic equation (8.4) has the form

$$0 = a^2|\xi|^2 \quad \Longleftrightarrow \quad \xi = 0. \tag{8.9}$$

It does not have the roots $\tau(\xi)$ for $\xi \neq 0$, hence, the heat equation is not hyperbolic in t (it is called *parabolic* instead; see Appendix A). The cone Q consists of vectors parallel to the axis Ot:

$$Q = \{(\tau,0,0,0)\},$$

where $\tau \in \mathbb{R}$ is arbitrary. The characteristic planes are given by equations $t = \text{const}$ and are orthogonal to the axis Ot (Fig. 8.3).

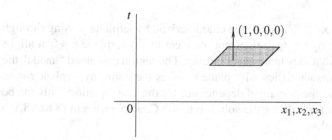

Fig. 8.3

Question 8.7. Is it true that equation (8.8) has solutions with discontinuities along the planes $t = \text{const}$?

Answer. No, it is not true. This is because equation (8.8) is not hyperbolic and because we neglected the term $\frac{\partial u}{\partial t}$ when writing the characteristic equation (8.9).

As the matter of fact, all solutions to the heat equation are smooth. On the other hand, it has solutions which are smooth on the characteristic planes $t = \text{const}$ but not analytic.

Example 8.8. The function $E(x,t) = \begin{cases} \frac{1}{(2\pi t)^{3/2}} e^{-\frac{|x|^2}{2t}}, & t > 0 \\ 0, & t \leq 0 \end{cases}$ where $x \in \mathbb{R}^3$

a. satisfies the heat equation (8.8) everywhere in $\mathbb{R}^4$, except at $(x,t) = (0,0)$;
b. For $t \neq 0$ or $x \neq 0$, it is smooth;
c. For each $x \neq 0$, this function is not analytic in t at $t = 0$.

Problem 8.9. Prove the three statements in Example 8.8.

Let us point out that if we remove the term $\frac{\partial u}{\partial t}$ from equation (8.8), the resulting equation $0 = \Delta u$, obviously, has solutions discontinuous on any given characteristic hyperplane $t = \text{const}$; for example, we could take functions of the form $u(x,t) \equiv f(t)$, where $f(t)$ is piecewise continuous. Therefore, contrary to the case of nondegenerate equations, properties of solutions to degenerate equations strongly depend on lower order terms.

Question 8.10. Is it possible to find the domain of dependence for a general equation (8.7) with the aid of characteristics, as in Section 4? In other words, does Conjecture 7.3 from Section 7 hold for this equation?

Answer. This conjecture is true indeed for a strictly hyperbolic equation (just as for the wave equation from Section 7). See [BJS79].

Remark 8.11. In a certain sense, Conjecture 7.3 is also true for the heat equation (8.8). Let us consider the Cauchy problem for equation (8.8) with the initial data

$$u\big|_{t=0} = \varphi(x). \tag{8.10}$$

For any point (x_0, t_0), $x_0 \in \mathbb{R}^3$, $t_0 > 0$ the characteristic hyperplane passing through it as unique and given by $t = t_0$. It does not intersect the hyperplane $t = 0$ at all, or instead one can think that they intersect at infinity. The region contained "inside" the intersections of the characteristics with plane $t = 0$ is the entire hyperplane $t = 0$. Indeed, this is precisely the domain of dependence for the heat equation. This can be seen from the Poisson formula for the solution to the Cauchy problem (8.8), (8.10) (see [Pet91, TS90, Vla84]):

$$u(x,t) = \frac{1}{(2\pi at)^{3/2}} \int_{\mathbb{R}^3} e^{-\frac{|x-y|^2}{2at}} \varphi(y)\, dy.$$

Thus,

‖ *The speed of propagation for the heat equation is infinite!*

Example 8.12. The Laplace equation is an example of the elliptic equation (see Appendix A):

$$\Delta u(x) \equiv \frac{\partial^2 u}{\partial x_1^2} + \frac{\partial^2 u}{\partial x_2^2} + \frac{\partial^2 u}{\partial x_3^2} = 0, \qquad x \in \mathbb{R}^3. \tag{8.11}$$

This equation can be obtained from the wave equation (7.1) and from the heat equation (8.8) when u does not depend on t. These are *stationary solutions*. Physically, they describe the stationary states of (7.1) or the limiting temperature distributions $t \to +\infty$ for solutions of equation (8.8) and are of particular interest in applications.

Let us find the plane wave solutions for (8.11):

$$u(x) = f(\langle \xi, x \rangle) = f(\xi_1 x_1 + \xi_2 x_2 + \xi_3 x_3), \qquad x \in \mathbb{R}^3. \tag{8.12}$$

Substituting into (8.11), we obtain, as above,

$$f''(\langle \xi, x \rangle)\xi_1^2 + f''(\langle \xi, x \rangle)\xi_2^2 + f''(\langle \xi, x \rangle)\xi_3^2 = 0, \tag{8.13}$$

getting the characteristic equation

$$\xi_1^2 + \xi_2^2 + \xi_3^2 = 0. \tag{8.14}$$

It follows that $\xi_1 = \xi_2 = \xi_3 = 0$.

Conclusion. Equation (8.11) is not hyperbolic (in either variable).

Question 8.13. Does this mean that the Laplace equation has no solutions similar to the plane waves?

Answer. No, it does not. Let us take a complex solution to (8.14); for example,

$$\xi_1 = i\sqrt{\xi_2^2 + \xi_3^2}, \qquad (\xi_2, \xi_3) \in \mathbb{R}^2.$$

The function $f(z)$ in (8.12) should then be defined for complex values of z. More-over, in the first term in (8.13), $f''(\langle \xi, x \rangle)$ is the derivative of f in the direction of the imaginary axis, while in the second and the third it is the derivative in the di-rection of the real axis! Therefore, to cancel f'' out of (8.13) and to get (8.14), we need $f(z)$ to have the same values of the derivatives in the directions of the real and imaginary axes at each point. But, as known from the theory of functions of com-plex variable, this means that $f(z)$ should be analytic in the entire complex plane (the *entire function*)! Consequently, $u(x) = f(\langle \xi, x \rangle)$ is a real analytic function of real variables x_1, x_2, x_3 and can not be discontinuous. For example,

$$u(x) = \langle \xi, x \rangle^3 = (x_1 i \sqrt{\xi_2^2 + \xi_3^2} + \xi_2 x_2 + \xi_3 x_3)^3.$$

Corollary 8.14. All the plane wave solutions to the Laplace equation (8.11) are real analytic and, consequently, smooth.

Remark 8.15. It turns out that *all* the solutions to the Laplace equation are real-analytic [Pet91].

The shock waves and the Cherenkov radiation

Let us consider the electromagnetic field of a charge which moves steadily in a certain medium. If its velocity is equal to v and it moves in the positive direction of the axis Ox_1, then its electromagnetic field is described by four potentials, each of them having the form

$$\varphi(x,t) = u(x_1 - vt, x_2, x_3) \tag{8.15}$$

and satisfies the wave equation (7.1) everywhere away from the point $(x_1 - vt, 0, 0) \equiv 0$ where the charge is located. The value of a in (7.1) is given by $a = c_b$, where c_b is the speed of light in the medium. Let us point out that $c_b < c$, where c is the speed of light in the vacuum, while v could be greater or smaller than c_b (but less than c).

Substituting (8.15) into (7.1), we get the equation

$$v^2 \frac{\partial^2 u}{\partial x_1^2}(x_1 - vt, x_2, x_3) = c_b^2 \left(\frac{\partial^2 u}{\partial x_1^2} + \frac{\partial^2 u}{\partial x_2^2} + \frac{\partial^2 u}{\partial x_3^2} \right), \quad x \neq x(t),$$

from where, denoting $x_1 - vt = y_1$, we get the following equation on $u(y_1, x_2, x_3)$:

$$(c_b^2 - v^2) \frac{\partial^2 u}{\partial y_1^2} + c_b^2 \left(\frac{\partial^2 u}{\partial x_2^2} + \frac{\partial^2 u}{\partial x_3^2} \right) = 0, \quad (y_1, x_2, x_3) \neq 0. \tag{8.16}$$

The characteristic equation (8.4) which corresponds to (8.16) is given by

$$(c_b^2 - v_b^2)\eta_1^2 + c_b^2(\eta_2^2 + \eta_3^2) = 0. \tag{8.17}$$

From here, we see that (*i*) when $v < c_b$, equation (8.16) does not have (real) characteristics (the same is true for the Laplace equation). It is of the elliptic type (see
Appendix A). It turns out that all its solutions are smooth, that is, the electromagnetic field does not have singularities for $x \neq x(t)$; (*ii*) when $v > c_b$, equation (8.16) is
hyperbolic in y_1, and, consequently, has discontinuous solutions similar to the plane
waves. For the characteristic cone Q represented by equation (8.17), as we know
from Section 7, there is a corresponding dual light cone K (see Fig. 8.4) described
by the equation

$$c_b^2 y_1^2 + (c_b^2 - v^2)(x_2^2 + x_3^2) = 0. \tag{8.18}$$

It turns out that the solution u is infinite in the part of the cone (8.18) where $y_1 < 0$.

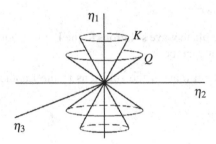

Fig. 8.4

From (8.18) we get the following equation of the surface of singularities of the
potential (8.15):

$$c_b^2 (x_1 - vt)^2 = (v^2 - c_b^2)(x_2^2 + x_3^2), \quad x_1 - vt < 0. \tag{8.19}$$

For each fixed t this surface in $\mathbb{R}^3$ is a cone with a vertex at the point $x(t)$ where the
charge is located (see Fig. 8.5). Along this surface the potentials and the field intensity are infinite, and the molecules of the medium at the points of the cone become

excited and emit the light. This is a physical phenomenon known as *Cherenkov radiation* (also known as Cherenkov – Vavilov radiation).

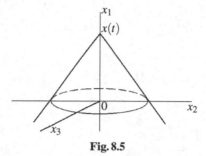

Fig. 8.5

The same situation arises when one is to find the sound generated by the body moving through the air: there are no pressure jumps if $v < c_{sound}$, but the jumps appear when $v > c_{sound}$. This is why behind the supersonic plane there is the shock wave located on the cone (8.19), that is, the pressure is discontinuous at the points of the cone (Fig. 8.6). We hear a bang when the pressure front passes our ear. The conic front of this shock wave is called the Mach cone.

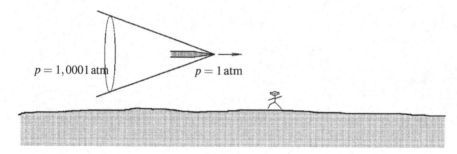

Fig. 8.6

Chapter 2
The Fourier method

9 Derivation of the heat equation

We consider a straight homogeneous metal rod of length l, cross-section S, and density ρ. We choose the axis x along the rod, and let $x = 0$ be the left end of the rod, so that $x = l$ is its right end. Denote by $u(x,t)$ the temperature of the rod at a point x at the moment $t > 0$. We assume that the cross-section is small, so that u depends only on x. It turns out that $u(x,t)$ satisfies the differential equation called the heat equation,

$$\frac{\partial u}{\partial t} = a^2 \frac{\partial^2 u}{\partial x^2}(x,t) + bf(x,t), \qquad (9.1)$$

where $f(x,t)$ is the density of the external heat source at the point x at the moment t. This means that the piece $[x, x + \Delta x]$ of the rod during the time interval from t until $t + \Delta t$ receives from the outside the amount of heat equal to

$$Q_{external} = f(x,t)\Delta x \Delta t. \qquad (9.2)$$

Let us derive (9.1). To do this, we write the equation of the heat balance for the piece of the rod $[x, x + \Delta x]$ as the time changes from t to $t + \Delta t$:

$$cm\Delta T = Q. \qquad (9.3)$$

Here c is the specific heat capacity of the material, $m = \rho S \Delta x$ is the mass of the piece, and ΔT is the temperature increase:

$$\Delta T \approx u(x, t + \Delta t) - u(x,t). \qquad (9.4)$$

Q is the total amount of heat received by the piece:

$$Q = Q_{external} + Q_l + Q_r, \qquad (9.5)$$

Alexander Komech and Andrew Komech, *Principles of Partial Differential Equations*, Problem Books in Mathematics, DOI 10.2007/978-1-4419-1096-7_2, © Springer Science + Business Media, LLC 2009

where $Q_{external}$ is the heat received from the external sources, Q_l is the amount of heat received from the left (that is, through the section of the rod at the point x), while Q_r is the amount of heat received from the right (that is, through the section of the rod at the point $x + \Delta x$). See Fig. 9.1.

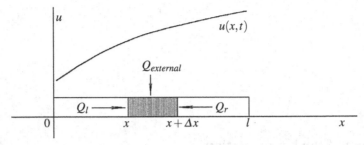

Fig. 9.1

According to the Fourier law of heating,

$$Q_l = -\lambda S \frac{\partial u}{\partial x}(x,t)\Delta t, \qquad Q_r = \lambda S \frac{\partial u}{\partial x}(x + \Delta x, t)\Delta t, \qquad (9.6)$$

where λ is the heat transfer coefficient and S is the cross-section area of the rod. The relation (9.6) means that the rate of the heat transfer through the cross-section of the rod at the point x is proportional to the rate of change of the temperature, $\frac{\partial u}{\partial x}(x,t)$. Signs in (9.6) are chosen so that the heat is transferred from warmer bodies to cooler ones (the second law of thermodynamics). For example, for $u(x,t)$ on Fig. 9.1, $Q_l < 0$, $Q_r > 0$, while $\frac{\partial u}{\partial x} > 0$ everywhere, hence the signs in the left- and right-hand sides of (9.6) coincide.

Substituting (9.6) and (9.2) into (9.5), and then (9.5) and (9.4) into (9.3), we get

$$c\rho S\Delta x\left(u(x,t+\Delta t) - u(x,t)\right) \approx f(x,t)\Delta x\Delta t + \lambda S\left(\frac{\partial u}{\partial x}(x+\Delta x,t) - \frac{\partial u}{\partial x}(x,t)\right)\Delta t.$$

From here, dividing by $\Delta x\Delta t$ and considering the limit $\Delta x \to 0$ and $\Delta t \to 0$, we get

$$c\rho S\frac{\partial u}{\partial t} = \lambda S\frac{\partial^2 u}{\partial x^2} + f(x,t). \qquad (9.7)$$

Then (9.1) follows, with the values of the constants being $a^2 = \frac{\lambda}{c\rho}$ and $b = \frac{1}{c\rho S}$.

10 Mixed problem for the heat equation

Here we will describe the basic idea of the Fourier method.

To determine the temperature of the rod, besides equation (9.1), one needs to specify the initial temperature

$$u(x,0) = \varphi(x), \qquad 0 < x < l \tag{10.1}$$

and the boundary conditions. For example, if the ends of the rod are submerged into the melting ice, then their temperature will be equal to zero ($0°\,$C):

$$u(0,t) = 0, \qquad u(l,t) = 0, \qquad t > 0. \tag{10.2}$$

The problem (9.1), (10.1), (10.2) is called the mixed problem for the heat equation.

For simplicity, we first assume that $f(x,t) \equiv 0$. The general case with the nonhomogeneity $f(x,t) \neq 0$ is considered in Section 15 below. Let us write the problem (9.1), (10.1), (10.2) (with $f \equiv 0$) in the operator form:

$$\begin{cases} \frac{d}{dt}\hat{u} = a^2 A \hat{u}(t), & t > 0; \\ \hat{u}(0) = \hat{\varphi}. \end{cases} \tag{10.3}$$

Here $A = \frac{d^2}{dx^2}$, $\hat{u}(t) \equiv u(x,t)$, and $\hat{\varphi} \equiv \varphi(x)$. As it follows from the boundary conditions (10.2), $\hat{u}(t) \in C_0^2[0,l]$ for all $t > 0$, where

$$C_0^2[0,l] \equiv \{u(x) \in C^2[0,l] \;:\; u(0) = u(l) = 0\}.$$

Thus, we consider the operator $A = \frac{d^2}{dx^2}$ on the domain $D(A) = C_0^2[0,l]$.

The idea of the Fourier method is to try to find a solution to the problem (10.3) in the form of the sum of particular solutions of the form $T(t)X(x)$. Let us illustrate this idea on an example of the system of n ordinary differential equations with n unknown functions, also written in the vector form (10.3):

$$\begin{cases} \frac{d}{dt}\hat{u}(t) = A\hat{u}(t), & \hat{u}(t) = (\hat{u}_1(t), \ldots, \hat{u}_n(t)) \in \mathbb{R}^n, \qquad t > 0; \\ \hat{u}(0) = \hat{\varphi} = (\hat{\varphi}_1, \ldots, \hat{\varphi}_n) \in \mathbb{R}^n, \end{cases} \tag{10.4}$$

where A is a matrix of size $n \times n$. Assume that there is a basis of the eigenvectors $e_1, \ldots, e_n$ of the matrix A, with the eigenvalues λ_k:

$$Ae_k = \lambda_k e_k, \qquad k = 1, \ldots, n. \tag{10.5}$$

Then the solution $\hat{u}(t)$ we are looking for, as well as the initial vector $\hat{\varphi}$, can be represented as

$$\hat{u}(t) = \sum_{k=1}^{n} T_k(t)e_k, \qquad \hat{\varphi} = \sum \varphi_k e_k.$$

Substituting into (10.4) we get

$$\sum_{k=1}^{n} \frac{dT_k(t)}{dt} e_k = \sum_{k=1}^{n} \lambda_k T_k(t) e_k, \qquad \sum_{k=1}^{n} T_k(0) e_k = \sum_{k=1}^{n} \varphi_k e_k,$$

hence

$$\frac{dT_k(t)}{dt} = \lambda_k T_k(t), \quad t > 0; \quad T_k(0) = \varphi_k.$$

We see that $T_k(t) = \varphi_k e^{\lambda_k t}$, and, therefore,

$$\hat{u}(t) = \sum_{k=1}^{n} \varphi_k e^{\lambda_k t} e_k. \tag{10.6}$$

In what follows we will obtain the analogs of formulas (10.5)–(10.6) for the operator $A = \frac{d^2}{dx^2}$.

11 The Sturm – Liouville problem

Let us find in $D(A) = C_0^2[0, l]$ the eigenfunctions $X_1(x), X_2(x), \ldots$ of the operator A:

$$\begin{cases} AX_k = \lambda_k X_k, & k \in \mathbb{N}; \\ X_k \in D(A), & X_k \neq 0. \end{cases} \tag{11.1}$$

The relation (11.1) means that

$$\begin{cases} X_k''(x) = \lambda_k X_k(x), & 0 < x < l; \\ X_k(0) = X_k(l) = 0, & X_k(x) \not\equiv 0. \end{cases} \tag{11.2}$$

Remark 11.1. We will show below in Section 13 that the solution to the problem (10.3) in the basis $X_1, \ldots, X_k, \ldots$ of the eigenfunctions of the operator A has the form analogous to (10.6):

$$u(x,t) = \sum_{k=1}^{\infty} e^{a^2 \lambda_k t} \varphi_k X_k(x), \tag{11.3}$$

where φ_k are the components of $\hat{\phi}$ in the basis $\{X_k : k \in \mathbb{N}\}$. Let us point out that in view of (11.1) each term in the series (11.3) satisfies the operator equation (10.3). Therefore any finite (partial) sum of this series also satisfies (10.3). The entire series (11.3) satisfies equation (10.3) if it allows termwise differentiation: once in t and twice in x. This is the case when the series converges sufficiently fast.

We introduce the notation

$$\langle u, v \rangle = \int_0^l u(x) v(x) \, dx \quad \text{for} \quad \forall u, v \in L^2[0, l].$$

Lemma 11.1. The operator $A = \frac{d^2}{dx^2}$ with the domain $D(A) = C_0^2[0,l]$ is symmetric and negative:

$$\left\langle \frac{d^2u}{dx^2}, v \right\rangle = \left\langle u, \frac{d^2v}{dx^2} \right\rangle, \qquad \forall u, v \in D(A), \tag{11.4}$$

$$\left\langle \frac{d^2u}{dx^2}, u \right\rangle < 0, \quad \forall u \in D(A), \qquad u(x) \not\equiv 0. \tag{11.5}$$

Proof. (*i*) The equality (11.4) means that

$$\int_0^l u''(x)\, v(x)\, dx = \int_0^l u(x)\, v''(x)\, dx. \tag{11.6}$$

To prove it, we integrate both sides of (11.6) by parts:

$$\int_0^l u''(x)\, v(x)\, dx = u' v \Big|_0^l \; - \int_0^l u'(x)\, v'(x)\, dx, \tag{11.7}$$

$$\int_0^l u(x)\, v''(x)\, dx = u v' \Big|_0^l \; - \int_0^l u'(x)\, v'(x)\, dx. \tag{11.8}$$

The boundary terms in the right-hand sides of (11.7) and (11.8) vanish since $v(0) = v(l) = 0$ and $u(0) = u(l) = 0$. Thus, the relation (11.6) is proved.
(*ii*) When $u = v$, it follows from (11.7) that

$$\left\langle \frac{d^2u}{dx^2}, u \right\rangle = \int_0^l u''(x)\, u(x)\, dx = - \int_0^l \left(u'(x) \right)^2 dx \le 0.$$

This proves (11.5). Indeed, if $\int_0^l \left(u'(x) \right)^2 dx = 0$, then $u'(x) \equiv 0$, $u(x) \equiv \text{const}$. But because of the boundary conditions $u(0) = u(l) = 0$ one concludes that $u(x) \equiv 0$, contradicting the condition $u(x) \not\equiv 0$ in (11.5).

Corollary 11.2. All the eigenvalues of the operator $A = d^2/dx^2$ are negative. Indeed, as it follows from (11.5),

$$0 > \left\langle \frac{d^2X_k}{dx^2}, X_k \right\rangle = \lambda_k \langle X_k, X_k \rangle.$$

The eigenfunctions X_k, X_n with different eigenvalues $\lambda_k \neq \lambda_n$ are orthogonal:

$$\int_0^l X_k(x)\, X_n(x)\, dx = 0.$$

Indeed, it follows from (11.4) that

$$\lambda_k \langle X_k, X_n \rangle = \langle AX_k, X_n \rangle = \langle X_k, AX_n \rangle = \lambda_n \langle X_k, X_n \rangle,$$

implying that $\langle X_k, X_n \rangle = 0$.

Solution of the Sturm – Liouville problem

From equation (11.2) we get

$$X_k(x) = A_k e^{\sqrt{\lambda_k}x} + B_k e^{-\sqrt{\lambda_k}x}. \tag{11.9}$$

Substituting this into the boundary conditions (11.2), we get

$$\begin{cases} A_k + B_k = 0, \\ A_k e^{\sqrt{\lambda_k}l} + B_k e^{-\sqrt{\lambda_k}l} = 0. \end{cases} \tag{11.10}$$

The matrix of this system should be degenerate, or else $A_k = B_k = 0$ and $X_k(x) \equiv 0$, contradicting (11.2). Thus, λ_k satisfy the characteristic equation

$$\det \begin{bmatrix} 1 & 1 \\ e^{\sqrt{\lambda_k}l} & e^{-\sqrt{\lambda_k}l} \end{bmatrix} = e^{-\sqrt{\lambda_k}l} - e^{\sqrt{\lambda_k}l} = 0.$$

It then follows that $e^{-\sqrt{\lambda_k}l} = e^{\sqrt{\lambda_k}l}$, hence $e^{2\sqrt{\lambda_k}l} = 1$. Therefore, $2\sqrt{\lambda_k}l = 2k\pi i$, $k \in \mathbb{Z}$, leading to

$$\sqrt{\lambda_k} = \frac{k\pi i}{l} \quad \Rightarrow \quad \lambda_k = -\left(\frac{k\pi}{l}\right)^2. \tag{11.11}$$

Here we may assume that $k \geq 0$. As one might have expected, $\lambda_k \leq 0$. Thus, the eigenvalues λ_k are found. Now let us find the eigenfunctions $X_k(x)$. For this, we take into account that the system (11.10) is degenerate. Therefore, these two equations are linearly dependent, and it suffices to consider only the first one: $B_k = -A_k$. In view of (11.11), we get:

$$X_k(x) = A_k\left(e^{\frac{k\pi i}{l}x} - e^{-\frac{k\pi i}{l}x}\right) = A_k 2i\sin\frac{k\pi x}{l}.$$

Here we applied the Euler formula

$$e^{i\varphi} - e^{-i\varphi} = (\cos\varphi + i\sin\varphi) - (\cos\varphi - i\sin\varphi) = 2i\sin\varphi.$$

Since the eigenfunctions X_k are defined up to a factor, we can finally set

$$X_k(x) = \sin\frac{k\pi x}{l}, \qquad k = 1, 2, \dots .$$

Here we can assume that $k > 0$, since for $k = 0$ we have $X_0(x) \equiv 0$.
Answer.

$$\lambda_k = -\left(\frac{k\pi}{l}\right)^2, \qquad X_k(x) = \sin\frac{k\pi x}{l}, \qquad k = 1, 2, \dots .$$

Properties of solutions to the Sturm – Liouville problem

Property 11.3. Completeness: $X_k(x)$ form a complete orthogonal set in $L^2(0,l)$ (this property is known from the theory of the Fourier series).

Property 11.4. Orthogonality:

$$\langle X_k, X_n \rangle = \int_0^l X_k(x) X_n(x)\, dx = 0 \quad \text{for} \quad k \neq n. \tag{11.12}$$

Property 11.5. Asymptotics: $\lambda_k \sim -k^2$ for $k \to \infty$. That is, there exists a limit

$$\lim_{k \to \infty} \frac{\lambda_k}{-k^2} > 0.$$

Problem 11.6. Check directly the orthogonality property (11.12) for X_k.

Solution. Since $k \neq n$,

$$\int_0^l \sin \frac{k\pi x}{l} \sin \frac{n\pi x}{l}\, dx = \frac{1}{2} \int_0^l \left(\cos \frac{(k-n)\pi x}{l} - \cos \frac{(k+n)\pi x}{l} \right) dx = 0.$$

Problem 11.7. Find the norm of X_k in $L^2(0,l)$.

Solution.

$$||X_k||^2 \equiv \int_0^l X_k^2(x)\, dx = \int_0^l \sin^2 \frac{k\pi x}{l}\, dx = \int_0^l \frac{1 - \cos \frac{2k\pi x}{l}}{2}\, dx = \frac{l}{2}. \tag{11.13}$$

Problem 11.8. Plot the graph of $X_k(x)$.

Solution. See Fig. 11.1.

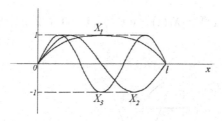

Fig. 11.1

Problem 11.9. Solve the Sturm – Liouville problem, that is, find the eigenfunctions of the operator $A \equiv \frac{d^2}{dx^2}$ on the interval $[0, l]$ for each of the boundary conditions:

$$X_k(0) = X_k'(l) = 0, \qquad\qquad (11.14)$$
$$X_k'(0) = X_k(l) = 0, \qquad\qquad (11.15)$$
$$X_k'(0) = X_k'(l) = 0. \qquad\qquad (11.16)$$

Answer.

For (11.14), $\lambda_k = -\left(\frac{(k+\frac{1}{2})\pi}{l}\right)^2$, $X_k(x) = \sin\frac{(k+\frac{1}{2})\pi x}{l}$, $k = 0, 1, 2, \dots$. See Fig. 11.2.

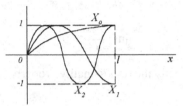

Fig. 11.2

For (11.15), $\lambda_k = -\left(\frac{(k+\frac{1}{2})\pi}{l}\right)^2$, $X_k(x) = \cos\frac{(k+\frac{1}{2})\pi x}{l}$, $k = 0, 1, 2, \dots$. See Fig. 11.3.

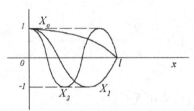

Fig. 11.3

For (11.16), $\lambda_k = -\left(\frac{k\pi}{l}\right)^2$, $X_k(x) = \cos\frac{k\pi x}{l}$, $k = 0, 1, 2, \dots$. See Fig. 11.4.

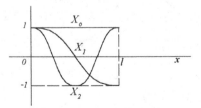

Fig. 11.4

One can also consider arbitrary boundary conditions of the form

$$\alpha_0 X_k'(0) + \beta_0 X_k(0) = 0, \qquad \alpha_1 X_k'(l) + \beta_1 X_k(l) = 0, \qquad (11.17)$$

where $\alpha_{0,1}$ and $\beta_{0,1}$ are real numbers such that

$$\alpha_0^2 + \beta_0^2 \neq 0, \qquad \alpha_1^2 + \beta_1^2 \neq 0.$$

Problem 11.10. Prove that the operator $\frac{d^2}{dx^2}$ with the boundary conditions (11.17) is symmetric.

Remark 11.11. The eigenfunctions and the eigenvalues corresponding to each of the boundary conditions (11.14), (11.15), and (11.16) possess all the properties 11.3, 11.4, and 11.5 (completeness, orthogonality, the asymptotics of the eigenvalues) of solutions to the Sturm – Liouville problem (11.1). See [Pet91, SD64, TS90, Vla84].

Multidimensional eigenvalue problem

Let us consider an arbitrary bounded region $\Omega \subset \mathbb{R}$ with a smooth boundary $\partial\Omega$ and the problem of finding the eigenfunctions of the Laplace operator in Ω with the Dirichlet boundary conditions:

$$\triangle X_k(x) = \lambda_k X_k(x), \qquad x \in \Omega,$$

$$X_k\Big|_{\partial\Omega} = 0. \qquad (11.18)$$

It turns out that its eigenfunctions corresponding to different λ_k are also orthogonal in $L^2(\Omega)$, while its eigenvalues λ_k are negative.

Problem 11.12. Prove that the Laplace operator with the boundary conditions (11.18) is symmetric and negative.

Problem 11.13. Prove that if instead of (11.18) one takes the Neumann boundary conditions,

$$\frac{\partial X_k}{\partial n}\Big|_{\partial\Omega} = 0,$$

where $\frac{\partial}{\partial n}$ stands for the derivative normal to $\partial\Omega$, then the Laplace operator is symmetric and non-positive, with $\lambda = 0$ the eigenvalue corresponding to the eigenfunction $X_0(x) \equiv 1$.

12 Eigenfunction expansions

As we already pointed out, the eigenfunctions $\sin\frac{k\pi x}{l}$, $k = 1, 2, \ldots$ form a complete orthogonal set in $L^2(0,l)$. Therefore they make up an orthogonal basis in $L^2(0,l)$ and, consequently, any function $\varphi(x) \in L^2(0,l)$ could be decomposed over this basis:

$$\varphi(x) = \sum_{k=1}^{\infty} \varphi_k X_k(x). \tag{12.1}$$

Let us find the formula for the coefficients φ_k. This is accomplished with the aid of the orthogonality conditions (11.12): we multiply (12.1) by $X_k(x)$ and integrate from 0 to l. Then we get

$$\int_0^l \varphi(x)X_n(x)\,dx = \sum_{k=1}^{\infty} \varphi_k \int_0^l X_k(x)X_n(x)\,dx = \varphi_n \int_0^l X_n^2(x)\,dx, \tag{12.2}$$

since all the terms in the summation in (12.2) with numbers $k \neq n$ are equal to zero! Termwise integration of the series in (12.2) is justified since the series in (12.1) converges in $L^2(0,l)$, while the scalar product in $L^2(0,l)$ is continuous in each of the two arguments.

Then from (12.2) we get the desired expression for the coefficients:

$$\varphi_n = \frac{\int_0^l \varphi(x)X_n(x)\,dx}{\int_0^l X_n^2(x)\,dx} = \frac{2}{l}\int_0^l \varphi(x)X_n(x)\,dx, \tag{12.3}$$

where we took into account that $\|X_n\|^2 = l/2$ by (11.13).

Problem 12.1. Find the conditions on the function $\varphi(x)$ so that the following is true:

(*i*) The series (12.1) converges uniformly on the interval $[0,l]$;
(*ii*) The series (12.1) is termwise differentiable two times.

Solution. (*i*) It is sufficient (but not necessary) that

$$\sum_{k=1}^{\infty} |\varphi_k| < \infty. \tag{12.4}$$

For this inequality to hold, it suffices to require that

$$\varphi(x) \in C^1[0,l], \qquad \varphi(0) = \varphi(l) = 0. \tag{12.5}$$

Let us derive (12.4) from (12.5). Integrating by parts, we get:

$$\varphi_k = \frac{2}{l} \int_0^l \varphi(x) \sin \frac{k\pi x}{l} \, dx = \frac{2}{l} \int_0^l \varphi(x) \frac{(-\cos \frac{k\pi x}{l})'}{\frac{k\pi}{l}} \, dx$$

$$= \frac{2}{k\pi} \left(-\varphi(x) \cos \frac{k\pi x}{l} \Big|_0^l + \int_0^l \varphi'(x) \cos \frac{k\pi x}{l} \, dx \right). \tag{12.6}$$

Above, the boundary terms are equal to zero due to the boundary conditions in (12.5). Therefore $\varphi_k = \frac{2}{k\pi} \varphi_k'$, where $\varphi_k' = \int_0^l \varphi'(x) \cos \frac{k\pi x}{l} \, dx$. But $\{\cos \frac{k\pi x}{l} : k \in \mathbb{N}\}$ is the orthogonal system in $L^2(0,l)$, with $\int_0^l \cos^2 \frac{k\pi x}{l} \, dx = \frac{l}{2}$, hence, due to the Bessel inequality,

$$\sum_{k=1}^{\infty} |\varphi_k'|^2 \leq \frac{2}{l} \int_0^l |\varphi'(x)|^2 \, dx < \infty. \tag{12.7}$$

Therefore, applying the Cauchy – Bunyakovsky inequality, we get:

$$\sum_{k=1}^{\infty} |\varphi_k| = \sum_{k=1}^{\infty} \left| \frac{2}{k\pi} \varphi_k' \right| \leq \left(\sum_{k=1}^{\infty} \left| \frac{2}{k\pi} \right|^2 \right)^{\frac{1}{2}} \left(\sum_{k=1}^{\infty} |\varphi_k'|^2 \right)^{\frac{1}{2}} < \infty. \tag{12.8}$$

(*ii*) For the series (12.1) to be twice differentiable, it suffices to have the series for $\varphi''(x)$ converge uniformly in x. The latter takes place if

$$\sum_{k=1}^{\infty} k^2 |\varphi_k| < \infty. \tag{12.9}$$

For this, we require that in addition to (12.5) we also have

$$\varphi(x) \in C^3[0,l] \quad \text{and} \quad \varphi''(0) = \varphi''(l) = 0. \tag{12.10}$$

Let us derive (12.9) from (12.10) and (12.5). For this, we remark that, due to (12.5) and (12.6),

$$\varphi_k = \frac{2}{k\pi} \int_0^l \varphi'(x) \cos \frac{k\pi x}{l} \, dx = \frac{2l}{(k\pi)^2} \left(\varphi'(x) \sin \frac{k\pi x}{l} \Big|_0^l - \int_0^l \varphi''(x) \sin \frac{k\pi x}{l} \, dx \right)$$

$$= \frac{2l^2}{(k\pi)^3} \left(\varphi''(x) \cos \frac{k\pi x}{l} \Big|_0^l - \int_0^l \varphi'''(x) \cos \frac{k\pi x}{l} \, dx \right).$$

The boundary terms vanished due to the boundary conditions (12.10) and due to $\sin \frac{k\pi x}{l}$ equal zero at $x = 0$ and $x = l$. Therefore, $\varphi_k = \frac{-2l^2}{(k\pi)^3} \varphi_k'''$, where $\varphi_k''' = \int_0^l \varphi'''(x) \cos \frac{k\pi x}{l} \, dx$. But $\varphi''' \in L^2(0,l)$; thus, by (12.7),

$$\sum_{k=1}^{\infty} |\varphi_k'''|^2 \leq \frac{2}{l} \int_0^l |\varphi'''(x)|^2 \, dx < \infty,$$

and, similarly to (12.8),

$$\sum_{k=1}^{\infty} k^2 |\varphi_k| = \sum_{k=1}^{\infty} k^2 \left| \frac{-2l}{(k\pi)^3} \varphi_k''' \right| \leq \frac{2l^2}{\pi^3} \sum_{k=1}^{\infty} \frac{1}{|k|} |\varphi_k'''| < \infty.$$

Problem 12.2. Show that for a function $\varphi(x) \in C^{(N)}[0,l]$ the estimates

$$|\varphi_k| \leq \frac{C}{|k|^N}, \qquad k = 1, 2, \ldots \tag{12.11}$$

are satisfied if and only if

$$\varphi(0) = \varphi(l) = 0, \quad \varphi''(0) = \varphi''(l) = 0, \quad \ldots, \quad \varphi^{2n}(0) = \varphi^{2n}(l) = 0 \tag{12.12}$$

for all $2n \leq N - 2, n = 0, 1, 2, \ldots$.

Let us point out that the boundary conditions (12.12) are satisfied, in particular, for all the eigenfunctions $\sin \frac{k\pi x}{l}$. On the other hand, under the condition (12.11), the series (12.1) is convergent on the interval $[0,l]$ uniformly together with its derivatives up to the order $N - 2$. Therefore, since the homogeneous boundary conditions (12.12) are satisfied for the eigenfunctions $\sin \frac{k\pi x}{l}$, it follows that the same boundary conditions are also satisfied for the sum of the series (12.1). This proves the necessity of conditions (12.12) for (12.11).

Remark 12.3. Similarly, let us consider the decomposition of the function $\varphi(x)$ over a system of eigenfunctions $X_k(x)$ which corresponds to each of the boundary conditions (11.14), (11.15), and (11.16). For estimate (12.11) for the Fourier coefficients φ_k of this decomposition to be true, it is necessary that $\varphi(x)$ satisfy the same homogeneous boundary conditions as the eigenfunctions $X_k(x)$ and their derivatives up to the order $N - 2$. When $\varphi \in C^{(N)}[0,l]$, it is easy to check that these conditions are not only necessary but also sufficient for (12.11).

Problem 12.4. Solve Problem 12.2 for the decomposition over the eigenfunctions of the Sturm – Liouville problem with each of the boundary conditions (11.14), (11.15), and (11.16).

Problem 12.5. Decompose over the set $\{\sin \frac{k\pi x}{l} : k \in \mathbb{N}\}$, the following functions:

a. $\varphi(x) \equiv 1, 0 < x < l$. See Fig. 12.1.

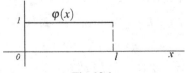

Fig. 12.1

Solution. $\varphi_k = \dfrac{2}{l} \displaystyle\int_0^l \sin \dfrac{k\pi x}{l}\, dx = -\dfrac{2}{l} \dfrac{\cos \frac{k\pi x}{l}}{\frac{k\pi}{l}} \Big|_0^l = \dfrac{2}{k\pi}\left[1 - (-1)^k\right].$

Let us point out that now the condition (12.4) is not satisfied. This is because $\varphi(x) \equiv 1$ is not equal to zero at the ends of the interval.

b. $\varphi(x) \equiv x,\ 0 < x < l$. See Fig. 12.2.

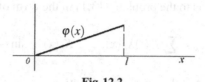

Fig. 12.2

Solution.

$$\varphi_k = \dfrac{2}{l} \int_0^l x \sin \dfrac{k\pi x}{l}\, dx = \dfrac{2}{l} \int_0^l x \dfrac{\left(-\cos \frac{k\pi x}{l}\right)'}{\frac{k\pi}{l}}\, dx = \ldots = -\dfrac{2}{k\pi} l(-1)^k.$$

Here $|\varphi_k| \sim \frac{1}{k}$ because $\varphi(l) \neq 0$ (see (12.11) and (12.12)).

c. $\varphi(x) = x(l - x)$. See Fig. 12.3.

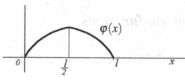

Fig. 12.3

Is it true that $\varphi_k = O(\frac{1}{k})$, or $O(\frac{1}{k^2})$, or $O(\frac{1}{k^3})$, …?

Problem 12.6. Decompose the functions $\varphi(x) = 1, x, x^2, x(l - x)$ over the eigen-functions of the Sturm – Liouville problem with each of the boundary conditions (11.14), (11.15), and (11.16). In each of these cases, find the asymptotics:

$$\varphi_k = O\left(\dfrac{1}{k}\right), \quad O\left(\dfrac{1}{k^2}\right), \quad \ldots.$$

Hint. Use Remark 12.3.

13 The Fourier method for the heat equation

So, let us solve the problem (10.3):

$$
\begin{cases}
\dfrac{\partial u}{\partial t} = a^2 \dfrac{\partial^2 u}{\partial x^2}, & u(0,t) = u(l,t) = 0, \qquad t > 0, \\
u(x,0) = \varphi(x), & 0 < x < l.
\end{cases} \tag{13.1}
$$

Let us look for a solution to the problem (13.1) in the form of the series

$$
u(x,t) = \sum_{k=1}^{\infty} T_k(t) X_k(x), \qquad X_k(x) = \sin \frac{k\pi x}{l}. \tag{13.2}
$$

Due to the completeness of the set of eigenfunctions

$$
\left\{ \sin \frac{k\pi x}{l} \ : \ k \in \mathbb{N} \right\} \tag{13.3}
$$

in $L^2(0,l)$, one can write in the form (13.2) any function $u(x,t)$ as long as $u(x,t) \in L^2(0,l)$ for each fixed t. The choice of the basis (13.3) is dictated by the boundary conditions which appear in (13.1). Namely, each term of the series (13.2) satisfies these boundary conditions since $\sin \frac{k\pi x}{l}$, $k \in \mathbb{N}$, satisfy the boundary conditions in (11.2).

To find the solution $u(x,t)$, it remains to determine *temporal functions* $T_k(t)$ (the functions $X_k(x)$ are called the *spatial functions*). $T_k(t)$ are found substituting the series (13.2) into the equation and the initial condition in (13.1).

Determining the temporal functions

A. We substitute the series (13.2) into equation (13.1): For $t > 0$,

$$
\sum_{k=1}^{\infty} T_k'(t) \sin \frac{k\pi x}{l} = -a^2 \sum_{k=1}^{\infty} T_k(t) \left(\frac{k\pi}{l} \right)^2 \sin \frac{k\pi x}{l}, \qquad 0 < x < l. \tag{13.4}
$$

Here we interchanged the operators of differentiation, $\frac{\partial}{\partial t}$ and $\frac{\partial^2}{\partial x^2}$, with the summation of the series. Below we will discuss why this interchange is allowed. The justification of the Fourier method is based on proving the validity of this interchange. In (13.4) we also used the identity

$$
\frac{\partial^2}{\partial x^2} \sin \frac{k\pi x}{l} = -\left(\frac{k\pi}{l} \right)^2 \sin \frac{k\pi x}{l}
$$

satisfied by the eigenfunctions of the Sturm – Liouville problem (11.1)–(11.2). Let us point out that the boundary conditions for the Sturm – Liouville problem have already been used.

Further, if the series in (13.4) converge in $L^2(0, l)$, then, due to the orthogonality of the basis $\{\sin \frac{k\pi x}{l} : k \in \mathbb{N}\}$, we get the following equations on the temporal functions $T_k(t)$:

$$T_k'(t) = -a^2 \left(\frac{k\pi}{l}\right)^2 T_k(t) = -\left(\frac{ak\pi}{l}\right)^2 T_k(t), \quad t > 0, \quad k = 1, 2, \ldots . \quad (13.5)$$

For each $k \in \mathbb{N}$, (13.5) is a homogeneous linear differential equation with constant coefficients. Let us write its characteristic equation:

$$\lambda = -\left(\frac{ak\pi}{l}\right)^2.$$

Then the general solution of (13.5) is given by

$$T_k(t) = C_k e^{-\left(\frac{ak\pi}{l}\right)^2 t}. \quad (13.6)$$

Substituting this expression into (13.2), we get

$$u(x, t) = \sum_{k=1}^{\infty} C_k e^{-\left(\frac{ak\pi}{l}\right)^2 t} \sin \frac{k\pi x}{l}. \quad (13.7)$$

B. The unknown constants C_k in (13.7) are found from the initial conditions. Namely, substituting the series (13.2) into the initial conditions in (13.1), we find:

$$\sum_{k=1}^{\infty} T_k(0) \sin \frac{k\pi x}{l} = \varphi(x), \quad 0 < x < l. \quad (13.8)$$

Hence, $T_k(0)$ coincide with the Fourier coefficients of the decomposition of the function $\varphi(x)$ over the set $\{\sin \frac{k\pi x}{l} : k \in \mathbb{N}\}$ (see (12.3)):

$$T_k(0) = \varphi_k \equiv \frac{2}{l} \int_0^l \varphi(x) \sin \frac{k\pi x}{l} dx. \quad (13.9)$$

Comparing with (13.6), we find

$$C_k = \varphi_k.$$

Thus, (13.7) takes the form

$$u(x, t) = \sum_{k=1}^{\infty} \varphi_k e^{-\left(\frac{ak\pi}{l}\right)^2 t} \sin \frac{k\pi x}{l}. \quad (13.10)$$

Justification of the Fourier method for the heat equation

Does the series (13.10) indeed represent the solution to the problem (13.1)?
A. For $t > 0$, the series (13.10) converge for each $x \in [0,l]$. For example, let

$$\varphi(x) \in L^2(0,l). \tag{13.11}$$

Then the series (13.8) converge in the same space $L^2(0,l)$. Indeed, from the Cauchy
– Bunyakovsky inequality,

$$|\varphi_k| \le \frac{2}{l}\int_0^l |\varphi(x)|\,dx \le \frac{2}{l}\left(\int_0^l dx\right)^{\frac{1}{2}}\left(\int_0^l \varphi^2(x)\,dx\right)^{\frac{1}{2}} \le \text{const}.$$

Therefore, the series (13.10) for each fixed $t > 0$ is dominated by the series

$$\text{const}\sum_{k=1}^\infty e^{-(\frac{ak\pi}{l})^2 t} = \text{const}\sum_{k=1}^\infty e^{-\varepsilon k^2},$$

where $\varepsilon = (\frac{a\pi}{l})^2 t > 0$, which in turn is dominated by the convergent geometric
series. Hence, according to the Weierstrass Theorem, the functional series (13.10)
converges uniformly on $[0,l]$ for $\forall t > 0$ to a function which is continuous in x.

Corollary 13.1. The series (13.10) satisfies the boundary conditions (10.2).

B. The series (13.10) is a differentiable function in $x \in [0,l]$ for any $t > 0$. Indeed,
according to the theorem about the termwise differentiation of a series,

$$\frac{\partial u}{\partial x}(x,t) = \sum_{k=1}^\infty \varphi_k e^{-(\frac{ak\pi}{l})^2 t}\left(-\cos\frac{k\pi x}{l}\right)\frac{k\pi}{l}, \tag{13.12}$$

as long as the series in the right-hand side converges uniformly in x on $[0,l]$. The
last condition is satisfied for any $t > 0$ since the series (13.12) is dominated by the
convergent series

$$\text{const}\frac{\pi}{l}\sum_{k=1}^\infty k e^{-\varepsilon k^2} < \infty.$$

C. The series (13.10) has derivatives in x and in t of all orders for $t > 0$. This is
proved similarly to **B**.

Corollary 13.2. All termwise differentiations of series in (13.4) are justified, hence
the series (13.10) satisfies the heat equation (9.1).

Finally, for $t = 0$ the series (13.10) satisfies the initial condition (10.1) in view of
(13.8) and (13.9) in the following sense (prove this!):

$$\|u(x,t) - \varphi(x)\|_{L^2(0,l)} \to 0 \quad \text{for} \quad t \to 0+.$$

Remark 13.3. The condition (13.11) allows the function $\varphi(x)$ to have disconti-nuities: For example, let $\varphi(x) \equiv 0$ for $x < \frac{l}{2}$, $u(x) \equiv 1$ for $x < \frac{l}{2}$. Then the func-tion $u(x,0) = \varphi(x)$ will be discontinuous. At the same time, the solution $u(x,t)$ for any $t > 0$ will be a smooth function on $[0,l]$! One says that the heat equation (9.1) "smoothens" the initial data.

Problem 13.4. Find the solution to the mixed problem

$$\begin{cases} \frac{\partial u}{\partial t} = 9\frac{\partial^2 u}{\partial x^2}(x,t), & 0 < x < 5, \quad t > 0; \\ u(0,t) = u(5,t) = 0; \\ u(x,0) = 1. \end{cases}$$

Solution. According to (13.10),

$$u(x,t) = \sum_{k=1}^{\infty} \varphi_k e^{-(\frac{3k\pi}{5})^2 t} \sin\frac{k\pi x}{5}, \tag{13.13}$$

where φ_k are found using (13.9):

$$\varphi_k = \frac{2}{5}\int_0^5 \sin\frac{k\pi x}{5}\,dx = \frac{2}{k\pi}[1-(-1)^k].$$

Problem 13.5. Find the limit of the solution (13.13) for $t \to \infty$.

Solution.

$$\lim_{t\to+\infty} u(x,t) = \lim_{t\to+\infty} \sum_{k=1}^{\infty} \varphi_k e^{-(\frac{3k\pi}{5})^2 t}\sin\frac{k\pi x}{5}$$
$$= \sum_{k=1}^{\infty} \varphi_k \lim_{t\to+\infty} e^{-(\frac{3k\pi}{5})^2 t}\sin\frac{k\pi x}{5} = \sum_{k=1}^{\infty} 0 = 0. \tag{13.14}$$

Problem 13.6. Justify the interchange of taking the limit and the summation in (13.14).

Problem 13.7. Find the solution to the mixed problem

$$\begin{cases} u_t(x,t) = 4u_{xx}(x,t), & 0 < x < 3, \quad t > 0; \\ u(0,t) = 0, & u_x(3,t) = 0; \\ u(x,0) = x. \end{cases} \tag{13.15}$$

Solution. Here the solution should be decomposed over the eigenfunctions of the Sturm – Liouville problem (11.14) (see Fig. 11.2):

$$u(x,t) = \sum_{k=0}^{\infty} T_k(t)\sin\frac{(k+\frac{1}{2})\pi x}{3}. \tag{13.16}$$

Substituting this series into (13.15), we obtain

$$\sum_{k=0}^{\infty} T_k'(t) \sin \frac{(k+\frac{1}{2})\pi x}{3} = 4\sum_{k=0}^{\infty} -\left(\frac{(k+\frac{1}{2})\pi}{3}\right)^2 T_k(t) \sin \frac{(k+\frac{1}{2})\pi x}{3}.$$

From this relation, for any $k = 0, 1, 2, \ldots$,

$$T_k'(t) = -\left(\frac{2(k+\frac{1}{2})\pi}{3}\right)^2 T_k(t) \quad \Rightarrow \quad T_k(t) = C_k e^{-\left(\frac{2(k+\frac{1}{2})\pi}{3}\right)^2 t}. \qquad (13.17)$$

Substituting (13.16) into the initial condition of the problem (13.15), we get

$$\sum_{k=0}^{\infty} T_k(0) \sin \frac{(k+\frac{1}{2})\pi x}{3} = x \quad \Longrightarrow$$

$$T_k(0) = \frac{2}{3} \int_0^3 x \sin \frac{(k+\frac{1}{2})\pi x}{3} \, dx$$

$$= \frac{2}{3} x \cdot \frac{-\cos \frac{(k+\frac{1}{2})\pi x}{3}}{\frac{(k+\frac{1}{2})\pi}{3}} \Big|_0^3 + \frac{2}{3} \int_0^3 \frac{\cos \frac{(k+\frac{1}{2})\pi x}{3}}{\frac{(k+\frac{1}{2})\pi}{3}} \, dx = 0 + \frac{2}{3} \frac{\sin \frac{(k+\frac{1}{2})\pi x}{3}}{\left(\frac{(k+\frac{1}{2})\pi}{3}\right)^2} \Big|_0^3$$

$$= \frac{\frac{2}{3} \sin(k+\frac{1}{2})\pi}{\left(\frac{(k+\frac{1}{2})\pi}{3}\right)^2} = \frac{2}{3} \frac{(-1)^k 9}{(k+\frac{1}{2})^2 \pi^2} = \frac{6(-1)^k}{(k+\frac{1}{2})^2 \pi^2}.$$

Since $C_k = T_k(0)$, we may now substitute $T_k(t)$ given by (13.17) into (13.16), getting

$$u(x,t) = \sum_{k=0}^{\infty} \frac{6(-1)^k}{(k+\frac{1}{2})^2 \pi^2} e^{-\frac{4\pi^2 (k+\frac{1}{2})^2 t}{9}} \sin \frac{(k+\frac{1}{2})\pi x}{3}.$$

Problem 13.8. Find the solution to the mixed problem

$$\begin{cases} u_t(x,t) = 16u_{xx}(x,t), & 0 < x < 3, \quad t > 0; \\ u_x(0,t) = u_x(3,t) = 0; \\ u(x,0) = x. \end{cases}$$

Problem 13.9. Find the limit $t \to \infty$ of the solution of the previous problem.

Answer.

$$\lim_{t \to \infty} u = \varphi_0 \equiv \frac{1}{3} \int_0^3 x \, dx = \frac{1}{3} \cdot \frac{9}{2} = \frac{3}{2}.$$

14 Mixed problem for the d'Alembert equation

Let us solve the mixed problem

$$\begin{cases} u_{tt}(x,t) = a^2 u_{xx}(x,t), & 0 < x < l, \quad t > 0; \\ u(0,t) = 0, \quad u(l,t) = 0; \\ u(x,0) = \varphi(x), \quad u_t(x,0) = \psi(x). \end{cases} \tag{14.1}$$

Similarly to (10.3), it is written in the operator form as

$$\begin{cases} \frac{\partial^2 \hat{u}}{\partial t^2}(t) = a^2 A \hat{u}(t), & t > 0; \\ \hat{u}(0) = \varphi, \quad \frac{\partial \hat{u}}{\partial t}(0) = \psi. \end{cases}$$

Solution of the problem (14.1)

We will look for the solution in the form of the series (13.2):

$$u(x,t) = \sum_{k=1}^{\infty} T_k(t) \sin \frac{k\pi x}{l}. \tag{14.2}$$

A. Substituting (14.2) into (14.1), we formally get

$$\sum_{k=1}^{\infty} T_k''(t) \sin \frac{k\pi x}{l} = a^2 \sum_{k=1}^{\infty} -\left(\frac{k\pi}{l}\right)^2 T_k(t) \sin \frac{k\pi x}{l}.$$

From here, as long as these series converge in $L^2(0,l)$, we find the equations on the temporal functions (compare with (13.5)):

$$T_k''(t) = -\left(\frac{ak\pi}{l}\right)^2 T_k(t), \quad \forall k = 1, 2, \ldots. \tag{14.3}$$

The general solution is (compare with (13.6)):

$$T_k(t) = A_k \cos \frac{ak\pi}{l} t + B_k \sin \frac{ak\pi}{l} t. \tag{14.4}$$

B. The unknown constants A_k and B_k are found from the initial conditions in (14.1):

$$\begin{cases} u(x,0) = \sum_{k=1}^{\infty} T_k(0) \sin \frac{k\pi x}{l} = \varphi(x) \implies T_k(0) = \varphi_k \quad \text{(see (13.9))}, \\ u_t(x,0) = \sum_{k=1}^{\infty} T_k'(0) \sin \frac{k\pi x}{l} = \psi(x) \implies T_k'(0) = \psi_k \equiv \frac{2}{l} \int_0^l \psi_k(x) \sin \frac{k\pi x}{l} dx. \end{cases}$$

Substituting (14.4) in the above relation, we find:

$$T_k(0) = A_k = \varphi_k,$$

$$T_k'(0) = B_k \frac{ak\pi}{l} = \psi_k \quad \Rightarrow \quad B_k = \frac{\psi_k}{\left(\frac{ak\pi}{l}\right)}. \tag{14.5}$$

Therefore, according to (14.4),

$$T_k(t) = \varphi_k \cos \frac{ak\pi}{l} t + \frac{\psi_k}{\left(\frac{ak\pi}{l}\right)} \sin \frac{ak\pi}{l} t.$$

Finally, substituting (14.5) into (14.2), we obtain:

$$u(x,t) = \sum_{k=1}^{\infty} \left(\varphi_k \cos \frac{ak\pi}{l} t + \frac{\psi_k}{\left(\frac{ak\pi}{l}\right)} \sin \frac{ak\pi}{l} t \right) \sin \frac{k\pi x}{l} t. \tag{14.6}$$

Question 14.1. While deriving (14.3), we again interchanged differentiation in x and t with the summation. Is this justified?

Justification of the Fourier method for the d'Alembert equation

A. Does the series (14.6) converge? It is dominated by the series

$$\text{const} \sum_{k=1}^{\infty} \left(|\varphi_k| + \frac{|\psi_k|}{k} \right).$$

For the convergence of this series, it suffices that

$$\begin{cases} \varphi(x) \in C^1[0,l], & \varphi(0) = \varphi(l) = 0; \\ \psi(x) \in C[0,l]. \end{cases}$$

This is proved similarly to the derivation of (12.4) from (12.5).
B. We need to be able to differentiate the series (14.6) twice in x and in t. For this, the convergence of the following series suffices:

$$\sum_{k=1}^{\infty} \left(k^2 |\varphi_k| + k |\psi_k| \right) < \infty. \tag{14.7}$$

For the convergence of this series, it is sufficient to have

$$\begin{cases} \varphi(x) \in C^3[0,l], & \varphi(0) = \varphi(l) = 0, & \varphi''(0) = \varphi''(l) = 0; \\ \psi(x) \in C^2[0,l], & \psi(0) = \psi(l) = 0. \end{cases} \tag{14.8}$$

This is proved analogously to the derivation of (12.9) from (12.10).

Conclusion. The series (14.6) is a solution to the problem (14.1) if the functions φ and ψ satisfy the conditions (14.8).

Remark 14.2. More precise (less restrictive) conditions on φ, ψ are given in terms of the Sobolev spaces (see Section 26 below).

Problem 14.3. Find the solution of the mixed problem

$$\begin{cases} u_t = 9u_{xx}(x,t), & 0 < x < 4, \quad t > 0; \\ u_x(0,t) = 0, & u(4,t) = 0; \\ u(x,0) = 0, & u_t(x,0) = 16 - x^2. \end{cases} \qquad (14.9)$$

Solution. One needs to decompose the solution over the eigenfunctions of the Sturm – Liouville problem (11.15) (see Fig. 11.3):

$$u(x,t) = \sum_{k=0}^{\infty} T_k(t) \cos \frac{(k+\frac{1}{2})\pi x}{4}.$$

Substitution into (14.9) gives, similarly to (14.3),

$$T_k''(t) = -9 \left(\frac{(k+\frac{1}{2})\pi}{4} \right)^2 T_k(t). \qquad (14.10)$$

The initial conditions in (14.9) give

$$\begin{cases} T_k(0) = \varphi_k = 0, \\ T_k'(0) = \psi_k \equiv \dfrac{2}{4} \displaystyle\int_0^4 (16-x^2) \cos \dfrac{(k+\frac{1}{2})\pi x}{4} dx = \dfrac{4^3(-1)^k}{(k+\frac{1}{2})^3 \pi^3}. \end{cases} \qquad (14.11)$$

Let us point out that here $\varphi_k \equiv 0$, while $\psi(x)$ satisfies conditions similar to (14.8): $\psi(x) \equiv 16 - x^2 \in C^2[0,4]$; $\psi'(0) = \psi(4) = 0$, that is, $\psi(x)$ satisfies the same homogeneous boundary conditions as the eigenfunctions $X_k(x) = \cos \frac{(k+\frac{1}{2})\pi x}{4}$ do, and $|\psi_k| \le C/k^3$ due to Remark 12.3. Therefore, the estimate (14.7) takes place.
From (14.10) and (14.11) we find, similarly to (14.4) and (14.5):

$$T_k(t) = \frac{\psi_k \sin \frac{3(k+\frac{1}{2})\pi t}{4}}{\frac{3(k+\frac{1}{2})\pi}{4}}.$$

Answer.

$$u(x,t) = \sum_{k=1}^{\infty} \frac{256(-1)^k}{3\left((k+\frac{1}{2})\pi\right)^4} \sin \frac{3(k+\frac{1}{2})\pi t}{4} \cos \frac{(k+\frac{1}{2})\pi x}{4}.$$

15 The Fourier method for nonhomogeneous equations

The heat equation

A. Let us consider the mixed problem for the nonhomogeneous heat equation with the homogeneous boundary conditions (nonhomogeneous boundary conditions in Section 16 below will be the next step in developing the Fourier method):

$$\begin{cases} \frac{\partial u}{\partial t} = a^2 \frac{\partial^2 u}{\partial x^2} + f(x,t), & 0 < x < l; \\ u(0,t) = 0, & u(l,t) = 0; \\ u(x,0) = \varphi(x). \end{cases} \tag{15.1}$$

Again, we look a solution of this problem in the form (13.2):

$$u(x,t) = \sum_{k=1}^{\infty} T_k(t) \sin \frac{k\pi x}{l}. \tag{15.2}$$

The new step will be the decomposition of $f(x,t)$ over the eigenfunctions of the Sturm – Liouville problem:

$$f(x,t) = \sum_{k=1}^{\infty} f_k(t) \sin \frac{k\pi x}{l}; \qquad f_k(t) = \frac{2}{l} \int_0^l f(x,t) \sin \frac{k\pi x}{l} dx. \tag{15.3}$$

This decomposition is possible due to the completeness of the family of eigenfunctions $\sin \frac{k\pi x}{l}$, $k \in \mathbb{N}$, in the space $L^2(0,l)$ as long as $f(x,t) \in L^2(0,l)$ for each fixed $t > 0$.

B. For finding the temporal functions $T_k(t)$, we substitute decompositions (15.2), (15.3) into (15.1):

$$\sum_{k=1}^{\infty} T_k'(t) \sin \frac{k\pi x}{l} = -a^2 \sum_{k=1}^{\infty} \left(\frac{k\pi}{l}\right)^2 T_k(t) \sin \frac{k\pi x}{l} + \sum_{k=1}^{\infty} f_k(t) \sin \frac{k\pi x}{l}. \tag{15.4}$$

From here, due to the orthogonality of the family of eigenfunctions, we get

$$T_k'(t) = -\left(\frac{ak\pi}{l}\right)^2 T_k(t) + f_k(t), \qquad t > 0, \quad k = 1, 2, \dots. \tag{15.5}$$

Thus, the differential equation on the temporal functions is obtained. For the unique determination of these functions, one needs to take into account the initial condition from (15.1):

$$\sum_{k=1}^{\infty} T_k(0) \sin \frac{k\pi x}{l} = \varphi(x) \quad \Rightarrow \quad T_k(0) = \frac{2}{l} \int_0^l \varphi(x) \sin \frac{k\pi x}{l} dx.$$

Let us point out that the boundary conditions in (15.1) are automatically satisfied due to decomposition (15.2) (since they are satisfied for the eigenfunctions $\sin \frac{k\pi x}{l}$) as long as $T_k(t) = O\left(\frac{1}{k^2}\right)$.

C. Let us apply this scheme to particular problems.

Problem 15.1. Solve the mixed problem

$$
\begin{cases}
u_t = 16u_{xx} + 2, & 0 < x < 7, \quad t > 0; \\
u_x(0,t) = u(7,t) = 0; \\
u(x,0) = 0.
\end{cases}
\tag{15.6}
$$

Solution. As it follows from the boundary conditions, the solution should be decomposed over the eigenfunctions of the Sturm – Liouville problem (11.15) (see Fig. 11.3):

$$
u(x,t) = \sum_{k=0}^{\infty} T_k(t) \cos \frac{(k+\frac{1}{2})\pi x}{7}.
\tag{15.7}
$$

Substituting this series into (15.6), we get the equation similar to (15.5):

$$
T_k'(t) = -\left(\frac{4(k+\frac{1}{2})\pi}{7}\right)^2 T_k + f_k, \quad t > 0, \quad k = 1, 2, \ldots,
\tag{15.8}
$$

where

$$
f_k \equiv \frac{2}{7} \int_0^7 2\cos \frac{(k+\frac{1}{2})\pi x}{7} dx = \frac{4}{7} \frac{\sin \frac{(k+\frac{1}{2})\pi x}{7}}{\frac{(k+\frac{1}{2})\pi}{7}} \Bigg|_0^7 = 4\frac{(-1)^k}{(k+\frac{1}{2})\pi}.
\tag{15.9}
$$

As it follows from the initial condition of the problem,

$$
T_k(0) = 0.
\tag{15.10}
$$

Let us solve the problem (15.8), (15.10). The general solution to (15.8) has the form

$$
T_k(t) = T_k^0(t) + T_k^p(t),
\tag{15.11}
$$

where $T_k^0(t)$ is the general solution to the homogeneous equation,

$$
T_k^0(t) = C_k e^{-\left(\frac{4(k+\frac{1}{2})\pi}{7}\right)^2 t},
\tag{15.12}
$$

and a particular solution $T_k^p(t)$ to the nonhomogeneous equation (15.8) is a constant. Substituting $T_k^p(t) = A_k$ into (15.8), we get

$$
0 = -\left(\frac{4(k+\frac{1}{2})\pi}{7}\right)^2 A_k + f_k,
$$

$$A_k = \frac{49 f_k}{16\left((k+\frac{1}{2})\pi\right)^2} = \frac{49(-1)^k}{4\left((k+\frac{1}{2})\pi\right)^3}. \tag{15.13}$$

Substituting (15.12) and (15.13) into (15.11), we get

$$T_k(t) = C_k e^{-\left(\frac{4(k+\frac{1}{2})\pi}{7}\right)^2 t} + \frac{49(-1)^k}{4\left((k+\frac{1}{2})\pi\right)^3}. \tag{15.14}$$

Now we need to take into account the initial condition (15.10):

$$0 = C_k + \frac{49(-1)^k}{4\left((k+\frac{1}{2})\pi\right)^3} \quad \Rightarrow \quad C_k = -\frac{49(-1)^k}{4\left((k+\frac{1}{2})\pi\right)^3}.$$

Finally, substituting (15.14) into (15.7), we get

$$u(x,t) = \sum_{k=0}^{\infty} (-1)^k \frac{49}{4\left((k+\frac{1}{2})\pi\right)^3}\left(-e^{-\left(\frac{4(k+\frac{1}{2})\pi}{7}\right)^2 t} + 1\right)\cos\frac{(k+\frac{1}{2})\pi x}{7}. \tag{15.15}$$

Problem 15.2. Find the limit of the solution to the problem (15.6) as $t \to +\infty$.

Solution. Taking the limit $t \to \infty$ in each term in the series (15.15), we get (justify!)

$$u_\infty(x) \equiv \lim_{t \to +\infty} u(x,t) = \sum_{k=0}^{\infty} \frac{49(-1)^k}{4\left((k+\frac{1}{2})\pi\right)^3}\cos\frac{(k+\frac{1}{2})\pi x}{7}. \tag{15.16}$$

Let us compute the sum of this series. For this, we notice that

$$u_\infty'(x) = -\sum_{k=0}^{\infty} \frac{7}{4}\frac{(-1)^k}{\left((k+\frac{1}{2})\pi\right)^2}\sin\frac{(k+\frac{1}{2})\pi x}{7}, \tag{15.17}$$

$$u_\infty''(x) = -\sum_{k=0}^{\infty} \frac{(-1)^k}{4(k+\frac{1}{2})\pi}\cos\frac{(k+\frac{1}{2})\pi x}{7} = -\frac{1}{8} \tag{15.18}$$

where the last equality follows from decomposition (see (15.9))

$$2 = \sum_{k=0}^{\infty} \frac{4(-1)^k}{(k+\frac{1}{2})\pi}\cos\frac{(k+\frac{1}{2})\pi x}{7}.$$

Integrating twice the identity (15.18), we get

$$u_\infty(x) = \frac{1}{16}(-x^2 + C_1 x + C_2). \tag{15.19}$$

To find C_1 and C_2, we notice that due to (15.16) and (15.17)

$$u_\infty(7) = 0, \qquad u_\infty'(0) = 0.$$

Substituting the expression (15.19) into the above relations, we find $C_1 = 0$, $C_2 = 49$; hence, $u_\infty(x) = \frac{1}{16}(49 - x^2)$.

Remark 15.3. We could obtain u_∞ directly from (15.6), without using the non-stationary solution (15.15). To do so, we substitute u_t by 0 and solve the problem

$$\begin{cases} 0 = 16u_\infty''(x) + 2, & 0 < x < 7; \\ u_\infty'(0) = 0, & u_\infty(7) = 0. \end{cases} \tag{15.20}$$

Remark 15.4. The important property of the heat equation is that under stationary external conditions (that is, when the nonhomogeneous terms of the equation and the boundary conditions do not depend explicitly on t) the solution $u(x,t)$ stabilizes as $t \to +\infty$:

$$u(x,t) \to u_\infty(x), \qquad t \to +\infty. \tag{15.21}$$

The limit function $u_\infty(x)$ is the solution to the corresponding stationary problem. The only exception is the case of perfect insulation at both ends ($\frac{\partial u}{\partial x} = 0$ at $x = 0$ and $x = l$) and the nonhomogeneity is nonzero. In this case, there is no limit stationary state $u_\infty(x)$. If, for example, the nonhomogeneity in the equation is a positive constant (permanent heat influx), then the temperature growth is unbounded.

Problem 15.5. Find the limit as $t \to +\infty$ of the solution to the mixed problem

$$\begin{cases} u_t = 25u_{xx}(x,t) + 3x^2, & 0 < x < 6; \\ u(0,t) = 0, & u'(6,t) = 1; \\ u(x,0) = \sin x. \end{cases} \tag{15.22}$$

Solution. As we said above, we get from (15.22) and (15.21) the following boundary value problem for $u_\infty(x) = \lim_{t \to \infty} u(x,t)$:

$$\begin{cases} 0 = 25u_\infty''(x) + 3x^2, & 0 < x < 6; \\ u_\infty(0) = 0, & u_\infty'(6) = 1. \end{cases}$$

Integrating this equation, we get $u_\infty(x) = -\frac{x^4}{100} + C_1 x + C_2$. From the boundary conditions we get $C_2 = 0$, $-\frac{6^3}{25} + C_1 = 1$.

Answer. $u_\infty(x) = -\frac{x^4}{100} + \frac{241}{25}x$.

The wave equation

Let us consider the nonhomogeneous wave equation.

Problem 15.6. Solve the following mixed problem (where $\omega > 0$):

$$\begin{cases} u_{tt}(x,t) = 25u_{xx} + x(3-x)\sin \omega t, & 0 < x < 3, \quad t > 0; \\ u(0,t) = u(3,t) = 0; \\ u(x,0) = 0, & u_t(x,0) = 0. \end{cases} \tag{15.23}$$

Solution. **A.** In view of the boundary conditions in (15.23), we are looking for the solution u in form of the decomposition over the eigenfunctions of the Sturm – Liouville problem (11.1):

$$u(x,t) = \sum_{k=1}^{\infty} T_k(t)\sin\frac{k\pi x}{3}. \tag{15.24}$$

For this, the function $x(3-x)\sin\omega t$ in equation (15.23) is also decomposed into the series over the system $\sin\frac{k\pi x}{3}$:

$$x(3-x)\sin\omega t = \sum_{k=1}^{\infty} g_k \sin\frac{k\pi x}{3}\sin\omega t, \tag{15.25}$$

where $g_k = \frac{2}{3}\int_0^3 x(3-x)\sin\frac{k\pi x}{3}\,dx = \frac{36}{(k\pi)^3}\left(1-(-1)^k\right)$.

B. Finding the temporal functions $T_k(t)$. Substituting decomposition (15.24) and (15.25) into equation (15.23) and using the orthogonality of the family $\sin\frac{k\pi x}{3}$, we get, similarly to (14.3),

$$T_k''(t) = -\left(\frac{5k\pi}{3}\right)^2 T_k(t) + g_k\sin\omega t. \tag{15.26}$$

Substitution of the series (15.24) into the initial conditions (15.23) gives

$$T_k(0) = 0, \qquad T_k'(0) = 0. \tag{15.27}$$

The Cauchy problem (15.26)–(15.27) uniquely determines the temporal functions $T_k(t)$.

It is known that the general solution to equation (15.26) has the form

$$T_k(t) = T_k^0(t) + T_k^p(t), \tag{15.28}$$

where $T_k^0(t)$ is the general solution of the corresponding homogeneous equation

$$T_k^0(t) = A_k\cos\frac{5k\pi}{3}t + B_k\sin\frac{5k\pi}{3}t,$$

while $T_k^p(t)$ is a particular solution to the nonhomogeneous equation (15.26).

When finding a particular solution, one needs to distinguish two cases: the *resonant case* and the *non-resonant case*.

1. Non-resonant case: For all $k \in N$,

$$\omega \neq \frac{5k\pi}{3}. \tag{15.29}$$

Then $T_k^p(t)$ are to be looked for in the form

$$T_k^p(t) = C_k\sin\omega t.$$

Substitution into (15.26) gives

$$-\omega^2 C_k \sin \omega t = -\left(\frac{5k\pi}{3}\right)^2 C_k \sin \omega t + g_k \sin \omega t,$$

from where, in view of (15.29),

$$C_k = \frac{g_k}{\left(\frac{5k\pi}{3}\right)^2 - \omega^2}.$$

Then (15.28) takes the form

$$T_k(t) = A_k \cos \frac{5k\pi}{3} t + B_k \sin \frac{5k\pi}{3} t + \frac{g_k \sin \omega t}{\left(\frac{5k\pi}{3}\right)^2 - \omega^2}.$$

Finally, the initial conditions (15.27) yield

$$A_k = 0, \quad B_k \frac{5k\pi}{3} + \frac{g_k \omega}{\left(\frac{5k\pi}{3}\right)^2 - \omega^2} = 0 \quad \Rightarrow \quad B_k = -\frac{g_k \omega}{\frac{5k\pi}{3}\left(\left(\frac{5k\pi}{3}\right)^2 - \omega^2\right)}.$$

Thus, in the case when (15.29) is satisfied for all $k = 1, 2, \ldots$, we have

$$u(x,t) = \sum_{k=1}^{\infty} \frac{g_k}{\left(\frac{5k\pi}{3}\right)^2 - \omega^2} \left(-\frac{\omega}{\left(\frac{5k\pi}{3}\right)} \sin\left(\frac{5k\pi}{3}t\right) + \sin \omega t\right) \sin \frac{k\pi x}{3}. \qquad (15.30)$$

2. *Resonant case*: For some $m \in \mathbb{N}$,

$$\omega = \frac{5m\pi}{3}. \qquad (15.31)$$

In this case,

$$T_m^p(t) = t(C_m \cos \omega t + D_m \sin \omega t).$$

Taking $k = m$ and substituting into (15.26), we get

$$2(-C_m \omega \sin \omega t + D_m \omega \cos \omega t) + t(-C_m \omega^2 \cos \omega t - D_m \omega^2 \sin \omega t)$$
$$= -\left(\frac{5m\pi}{3}\right)^2 t(C_m \cos \omega t + D_m \sin \omega t) + g_m \sin \omega t. \qquad (15.32)$$

Here in the left-hand side we used the Leibniz formula for computing

$$\frac{d^2}{dt^2}\left[t(C_m \cos \omega t + D_m \sin \omega t)\right].$$

Taking into account (15.31) and collecting the terms in (15.32), we get

$$2(-C_m \omega \sin \omega t + D_m \omega \cos \omega t) = g_m \sin \omega t.$$

We compare the coefficients at $\cos \omega t$ and $\sin \omega t$ on the left and on the right:

$$2D_m\omega = 0, \qquad -2C_m\omega = g_m.$$

Since $\omega > 0$,

$$D_m = 0, \qquad C_m = -\frac{g_m}{2\omega}.$$

Thus,

$$T_m^p(t) = -t\frac{g_m}{2\omega}\cos\omega t.$$

Therefore

$$T_m(t) = A_m\cos\frac{5k\pi}{3}t + B_m\sin\frac{5k\pi}{3}t - t\frac{g_m}{2\omega}\cos\omega t.$$

Substituting into the initial conditions (15.27), we get

$$A_m = 0; \quad B_m\frac{5m\pi}{3} - \frac{g_m}{2\omega} = 0 \Longrightarrow B_m = \frac{3g_m}{10m\pi\omega}.$$

Therefore,

$$T_m^p(t) = \frac{3g_m}{10m\pi\omega}\sin\left(\frac{5k\pi}{3}t\right) - t\frac{g_m}{3}\cos\omega t.$$

Thus, if for some $m \in \mathbb{N}$ the condition (15.31) is satisfied, we get (compare with (15.30)):

$$u(x,t) = \sum_{k\in\mathbb{N}, k\neq m} \frac{g_k}{(\frac{5k\pi}{3})^2 - \omega^2}\left(-\frac{\omega}{(\frac{5k\pi}{3})}\sin\frac{5k\pi}{3}t + \sin\omega t\right)\sin\frac{k\pi x}{3}$$

$$+ \left(\frac{3g_m}{10m\pi\omega}\sin\frac{5m\pi}{3}t - t\frac{g_m}{2\omega}\cos\omega t\right)\sin\frac{m\pi x}{3}. \tag{15.33}$$

Remark 15.7. In the non-resonant case, all the terms in the series (15.30) are bounded functions of x, t, while in the resonant case (15.31) one of the terms in (15.33) is unbounded when $t \to +\infty$. Therefore, for large t, the solution will be represented mainly by the last term in (15.33). As t grows, the solution becomes unboundedly large. If it were the amplitude of a string, the string would break. As the matter of fact, when the solution becomes large, it is no longer described by the linear wave equation, and the formula (15.33) is no longer valid.

Problem 15.8. Find the solution to the mixed problem

$$\begin{cases} u_{tt}(x,t) = 16u_{xx} + \sin\frac{7\pi x}{10}, & 0 < x < 5, \quad t > 0; \\ u(0,t) = 0, & u_x(5,t) = 0; \\ u(0,x) = 0, & u_t(0,x) = 0. \end{cases}$$

16 The Fourier method for nonhomogeneous boundary conditions

Up to now, we were using the Fourier method only for problems with homogeneous boundary conditions. It turns out that the problem with nonhomogeneous boundary conditions is easily reduced to a problem with homogeneous boundary conditions.

The heat equation

Problem 16.1. Find the solution to the mixed problem

$$\begin{cases} u_t = 9u_{xx}, & 0 < x < 4, \quad t > 0; \\ u(0,t) = f(t), & u(4,t) = g(t); \\ u(x,0) = 0. \end{cases} \tag{16.1}$$

Solution. Let us find an auxiliary function $v(x,t)$ which satisfies the given boundary conditions:

$$v(0,t) = f(t), \qquad v(4,t) = g(t), \qquad t > 0.$$

Such a function can easily be found, for example, using a linear interpolation

$$v(x,t) = \frac{x}{4}g(t) + \frac{4-x}{4}f(t).$$

Denote $w = u - v$. Then w satisfies the homogeneous boundary conditions

$$w(0,t) = 0, \qquad w(4,t) = 0, \qquad t > 0. \tag{16.2}$$

Question 16.2. What equation and boundary conditions does the function w satisfy?

Answer. We substitute $u = w + v$ into (16.1); then

$$\begin{cases} w_t + v_t = 9(w_{xx} + v_{xx}), \\ w(x,0) + v(x,0) = 0, \end{cases}$$

leading to

$$\begin{cases} w_t = 9w_{xx} + 9(v_{xx} - v_t), \\ w(x,0) = -v(x,0). \end{cases}$$

Thus, unlike u, the function w satisfies the nonhomogeneous heat equation! But the boundary conditions (16.2) are now homogeneous, hence w could be found using the method of Section 15; then $u = w + v$ is the solution to the problem (16.1). Thus, we sent the nonhomogeneity from the boundary conditions into the differential equation (16.1) and into the initial condition.

The wave equation

Problem 16.3. Solve the mixed problem

$$\begin{cases} u_{tt} = 16u_{xx}, & 0 < x < 5, \quad t > 0; \\ u(0,t) = 0, & u_x(5,t) = \sin \omega t; \\ u(x,0) = 0, & u_t(x,0) = 0. \end{cases} \tag{16.3}$$

Solution. **A.** The auxiliary function

$$v(x,t) = x \sin \omega t$$

satisfies the specified boundary conditions. For $w \equiv u - v$, we have:

$$\begin{cases} w_{tt} = 16w_{xx} + \omega^2 x \sin \omega t, & 0 < x < 5, \quad t > 0; \\ w(0,t) = 0, & w_x(5,t) = 0; \\ w(x,0) = -v(x,0) = 0, & w_t(x,0) = -v_t(x,0) = -x\omega. \end{cases} \tag{16.4}$$

B. Following the method of Section 15, we are looking for w in the form

$$w(x,t) = \sum_{k=0}^{\infty} T_k(t) \sin \frac{(k+\frac{1}{2})\pi x}{5}. \tag{16.5}$$

For this, we expand the right-hand side of equation (16.4):

$$\omega^2 x \sin \omega t = \omega^2 \sin \omega t \sum_{k=0}^{\infty} x_k \sin \frac{(k+\frac{1}{2})\pi x}{5},$$

where

$$x_k = \frac{2}{5} \int_0^5 x \sin \frac{(k+\frac{1}{2})\pi x}{5} \, dx = -\frac{2}{5} \frac{5}{(k+\frac{1}{2})\pi} \int_0^5 x d\cos \frac{(k+\frac{1}{2})\pi x}{5}$$

$$= -\frac{5}{(k+\frac{1}{2})\pi} \left[x\cos \frac{(k+\frac{1}{2})\pi x}{5} \Big|_0^5 - \int_0^5 \cos \frac{(k+\frac{1}{2})\pi x}{5} \, dx \right]$$

$$= \frac{2 \cdot 5}{(k+\frac{1}{2})^2 \pi^2} \sin \frac{(k+\frac{1}{2})\pi x}{5} \Big|_0^5 = \frac{10}{(k+\frac{1}{2})^2 \pi^2} \cdot (-1)^k. \tag{16.6}$$

C. Substituting (16.5)–(16.6) into equation (16.4), we find the equations on the temporal functions $T_k(t)$:

$$T_k''(t) = -16 \left(\frac{(k+\frac{1}{2})\pi}{5} \right)^2 T_k(t) + \omega^2 x_k \sin \omega t, \qquad k = 0, 1, 2, \ldots. \tag{16.7}$$

From the initial conditions (16.4) we find $T_k(0) = 0$ and

$$T_k'(0) = \frac{2}{5} \int_0^5 (-\omega x) \sin \frac{(k+\frac{1}{2})\pi x}{5} \, dx = -\omega \frac{10 \cdot (-1)^k}{(k+\frac{1}{2})^2 \pi^2}. \qquad (16.8)$$

In the last equality, we took into account (16.6). The problem (16.7)–(16.8) could be solved in the same way as in Section 15. Again, two cases are possible: resonant and non-resonant.

Complete the solution of the problem (16.1).

Remark 16.4. For problems like (16.4) a condition analogous to (14.8) is not satisfied. Still, the new function $w(x,t)$ satisfies the initial and boundary conditions in the usual sense. It is only the first equation in (16.4) that is satisfied in the sense of distributions (see Section 26 below).

Problem 16.5. Find the resonance condition in the problem (16.3).

Answer. $\omega = \frac{4(m+\frac{1}{2})\pi x}{5}$ for some $m = 0, 1, 2, \dots$.

17 The Fourier method for the Laplace equation

Boundary value problems in a rectangle

A. Let us consider the boundary value problem in the rectangle $\Omega = [0,a] \times [0,b]$:

$$\begin{cases} \Delta u(x,y) \equiv \frac{\partial^2 u}{\partial x^2} + \frac{\partial^2 u}{\partial y^2} = 0, & 0 < x < a, \quad 0 < y < b; \\ u(0,y) = 0, & u(a,y) = 0; \\ u(x,0) = f(x), & u(x,b) = g(x). \end{cases} \qquad (17.1)$$

This is the boundary value problem, or *the Dirichlet problem*: the function u is given at the boundary of the considered region. See Fig. 17.1.

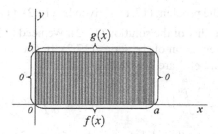

Fig. 17.1

Solution. The problem (17.1) can be solved by the method of Section 15, where the role of the variable t is now played by the variable y, as could be seen from comparing problems (17.1) and (15.1). We are looking for the solution in the following form:

$$u(x,y) = \sum_{k=1}^{\infty} Y_k(y) \sin \frac{k\pi x}{a}. \tag{17.2}$$

Then the boundary conditions at $x = 0$ and $x = a$ in (17.1) are automatically satisfied. We substitute (17.2) into equation (17.1). This gives equations on $Y_k(y)$:

$$-\left(\frac{k\pi}{a}\right)^2 Y_k(y) + Y_k''(y) = 0, \qquad 0 < y < b. \tag{17.3}$$

Substitution into the boundary conditions (17.1) at $y = 0$ and $y = b$ yields

$$\begin{cases} Y_k(0) = f_k \equiv \dfrac{2}{a} \displaystyle\int_0^a f(x) \sin \dfrac{k\pi x}{a}\, dx, \\[4mm] Y_k(b) = g_k \equiv \dfrac{2}{a} \displaystyle\int_0^a g(x) \sin \dfrac{k\pi x}{a}\, dx. \end{cases} \tag{17.4}$$

The general solution to equation (17.3) has the form

$$Y_k(y) = A_k e^{\frac{k\pi}{a} y} + B_k e^{-\frac{k\pi}{a} y}. \tag{17.5}$$

The constants A_k and B_k are found from the boundary conditions (17.4):

$$A_k + B_k = f_k, \qquad A_k e^{\frac{k\pi}{a} b} + B_k e^{-\frac{k\pi}{a} b} = g_k.$$

Solving this system, we find:

$$\begin{cases} A_k = \dfrac{1}{e^{\frac{k\pi}{a} b} - e^{-\frac{k\pi}{a} b}} \left(g_k - f_k e^{-\frac{k\pi}{a} b} \right), \\[4mm] B_k = \dfrac{1}{e^{\frac{k\pi}{a} b} - e^{-\frac{k\pi}{a} b}} \left(f_k e^{\frac{k\pi}{a} b} - g_k \right). \end{cases} \tag{17.6}$$

Thus, the solution of the problem (17.1) is given by (17.2), (17.5), and (17.6).

Let us check the validity of the solution (17.2). We need to justify the possibility of the termwise differentiation of the series (17.2). If $f(x)$ and $g(x)$ are integrable functions, then $f(x)$ and $g(x)$ are bounded:

$$|f_k| \le \frac{2}{a} \int_0^a |f(x)|\, dx, \qquad |g_k| \le \frac{2}{a} \int_0^a |g(x)|\, dx.$$

But then from (17.6) we see that $|A_k| \le \frac{c}{e^{\frac{k\pi}{a} b}}$, $|B_k| \le$ const. Therefore, it follows from (17.5) that $|Y_k(y)| \le c e^{-\frac{k\pi}{a}(b-y)} + c e^{-\frac{k\pi}{a} y}$. As a consequence, for $\varepsilon < y < b - \varepsilon$, with $\varepsilon > 0$ small, one has $|Y_k(y)| \le c e^{-\frac{k\pi}{a} \varepsilon}$, and the series (17.2) for these values of y is dominated by the convergent geometric series $\sum_{k=1}^{\infty} c e^{-\frac{k\pi}{a} \varepsilon}$. It is easy to see that the derivatives of the second order in x and in y of the series (17.2) are dominated by the series $\sum_{k=1}^{\infty} ck^2 e^{-\frac{k\pi}{a}\varepsilon}$, which is also convergent. In the same way one proceeds with the derivatives of any order in x and y.

Conclusion. Solution of the Dirichlet problem (17.1) is a smooth function inside the rectangle Ω. Let us assume that, as in (12.5), $f(x), g(x) \in C_0^2[0,a]$. Then, analogously to (12.4), $f_k, g_k = O(\frac{1}{k^2})$ and, consequently, $|Y_k(y)| \leq \frac{c}{k^2}, y \in [0,b]$. Therefore, the series (17.2) converges uniformly in the rectangle $\Omega = [0,a] \times [0,b]$, and its sum is a function which is continuous in this rectangle and satisfies boundary conditions in (17.1).

B. More general boundary value problem of the Dirichlet type in the rectangle,

$$\begin{cases} \Delta u(x,y) = 0, & 0 < x < a, \quad 0 < y < b; \\ u(0,y) = \varphi(y), & u(a,y) = \psi(y); \\ u(x,0) = f(x), & u(x,b) = g(x), \end{cases} \tag{17.7}$$

could be solved by decomposing the solution u into two terms:

$$u = u_1 + u_2. \tag{17.8}$$

Here u_1 solves the problem (17.1), while u_2 solves the problem

$$\begin{cases} \Delta u_2 = 0, & 0 < x < a, \quad 0 < y < b; \\ u_2(0,y) = \varphi(y), & u_2(a,y) = \psi(y); \\ u_2(x,0) = 0, & u_2(x,b) = 0. \end{cases}$$

This problem takes the same form as (17.1) if one interchanges x and y. Therefore u_2 should be tried in the form (compare with (17.2)):

$$u_2(x,y) = \sum_{k=1}^{\infty} X_k(x) \sin \frac{k\pi y}{b}. \tag{17.9}$$

If $f, g \in C_0^2[0,a]$, while $\varphi, \psi \in C_0^2[0,b]$, then, according to what we said above, u_1 and u_2, and, consequently, $u = u_1 + u_2$ are continuous functions in Ω which satisfy the required boundary conditions.

In the general case, for the continuity of $u(x,y)$ in Ω, the following compatibility conditions are obviously required:

$$f(0) = \varphi(0), \quad \varphi(b) = g(0), \quad g(a) = \psi(b), \quad \psi(0) = f(a). \tag{17.10}$$

Problem 17.1. Prove that the problem (17.7) has a solution continuous in Ω if $f, g \in C^2[0,a]$, $\varphi, \psi \in C^2[0,b]$, and the compatibility condition (17.10) is satisfied.

Hint. Try to find the solution to equation $\Delta v = 0$ in Ω which coincides with the boundary values given by functions f, g, φ, and ψ at the boundary of the region Ω. Then the difference $u - v$ could be found using decomposition (17.8) described above.

C. Now we consider the nonhomogeneous Laplace equation (*the Poisson equation*).

Problem 17.2. Solve the boundary value problem

$$\begin{cases} \triangle u(x,y) = x^2 y, & 0 < x < a, \quad 0 < y < b; \\ u(0,y) = 0, & u(a,y) = 0; \\ u(x,0) = 0, & \frac{\partial u}{\partial y}(x,b) = 0. \end{cases} \tag{17.11}$$

Let us point out that here at $x = 0$, $x = a$, and $y = 0$ one has the boundary value of the Dirichlet type, while at $y = b$ one has the boundary value of the Neumann type (that is, the derivative of the solution in the normal direction is specified).

Solution. Homogeneous boundary conditions at $x = 0$, and $x = a$ allow to write the solution in the form of the series over the eigenfunctions of the corresponding Sturm – Liouville problem:

$$u(x,y) = \sum_{k=1}^{\infty} Y_k(x) \sin \frac{k\pi y}{a}. \tag{17.12}$$

We also decompose over these functions the right-hand side:

$$x^2 y = y \sum_{k=1}^{\infty} g_k \sin \frac{k\pi y}{a}, \qquad g_k = \frac{2}{a} \int_0^a x^2 \sin \frac{k\pi y}{a} dx.$$

Substituting these decompositions into (17.11), we get for $\forall k = 1, 2, \ldots$

$$-\left(\frac{k\pi}{a}\right)^2 Y_k(y) + Y_k''(y) = y g_k, \quad 0 < y < b; \qquad Y_k(0) = 0, \quad Y_k'(b) = 0. \tag{17.13}$$

Then

$$Y_k(y) = A_k e^{\frac{k\pi x}{a}} + B_k e^{-\frac{k\pi x}{a}} + \frac{y g_k}{-\left(\frac{k\pi}{a}\right)^2}. \tag{17.14}$$

The constants A_k and B_k can be found after substituting this solution into the boundary conditions in (17.13):

$$A_k + B_k = 0, \qquad \frac{k\pi}{a} A_k e^{\frac{k\pi}{a} b} + \left(-\frac{k\pi}{a}\right) B_k e^{-\frac{k\pi}{a} b} + \frac{g_k}{-\left(\frac{k\pi}{a}\right)^2} = 0.$$

Answer. The solution is given by the formulas (17.12), (17.14).

Boundary value problems in annulus and in disc

A. Let us solve the boundary value problem of the Dirichlet type in the annulus between the circles of radii r_1 and r_2:

$$\begin{cases} \Delta u(x,y) = 0, & r_1^2 < x^2 + y^2 < r_2^2; \\ u|_{x^2+y^2=r_1^2} = f_1(\varphi), & u|_{x^2+y^2=r_2^2} = f_2(\varphi); \quad 0 \le \varphi \le 2\pi. \end{cases} \qquad (17.15)$$

Here f_1 and f_2 are given continuous functions of the angular variable φ.

Solution. Let us convert to polar coordinates $r = \sqrt{x^2 + y^2}$, $\varphi = \arctan \frac{y}{x}$.

Problem 17.3. Prove that in these coordinates the problem (17.15) takes the form

$$\begin{cases} \Delta u = \dfrac{\partial^2 u}{\partial r^2} + \dfrac{1}{r}\dfrac{\partial u}{\partial r} + \dfrac{1}{r^2}\dfrac{\partial^2 u}{\partial \varphi^2} = 0, & r_1 < r < r_2; \\ u|_{r=r_1} = f_1(\varphi), & u|_{r=r_2} = f_2(\varphi); \quad 0 \le \varphi \le 2\pi. \end{cases} \qquad (17.16)$$

This is a problem in a rectangle $[0, 2\pi] \times [r_1, r_2]$ (Fig. 17.2). The boundary conditions are given at the lower and at the upper sides of the rectangle.

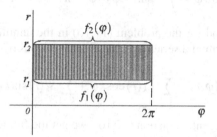

Fig. 17.2

Question 17.4. Are there boundary conditions at the left and right sides of the rectangle?

Answer. Yes, it is the periodicity condition in the variable φ:

$$u(0,r) = u(2\pi, r), \qquad \frac{\partial u}{\partial \varphi}(0,r) = \frac{\partial u}{\partial \varphi}(2\pi, r). \qquad (17.17)$$

This follows from the fact that the points with the polar coordinates $(0, r)$ and $(2\pi, r)$ are identical. Analogous periodicity conditions in φ also hold for all partial derivatives of u in r and φ.

Problem 17.5. Show that the conditions (17.17) together with equation (17.16) guarantee the periodicity in φ of all the derivatives of u in r and φ if $u(\varphi, r)$ is a smooth function in the rectangle $[0, 2\pi] \times [r_1, r_2]$.

The Sturm–Liouville problem which corresponds to the homogeneous boundary conditions (17.17) has the form

$$\begin{cases} \dfrac{\partial^2}{\partial\varphi^2}\Phi(\varphi) = \lambda\Phi(\varphi), & 0<\varphi<2\pi; \\ \Phi(0)=\Phi(2\pi), & \Phi'(0)=\Phi'(2\pi). \end{cases} \tag{17.18}$$

Solving this problem, we find:

$$\lambda_k = -k^2, \qquad \Phi_k(\varphi)=A_k\cos k\varphi+B_k\sin k\varphi, \qquad k=0,1,2,\dots.$$

Therefore, for each $k\neq 0$ there are two linearly independent eigenfunctions: $\cos k\varphi$ and $\sin k\varphi$, while for $k=0$ there is only one eigenfunction: $\Phi_0(\varphi)\equiv 1$. As it is known from the Fourier series theory, these eigenfunctions form a complete orthogonal set in $L^2(0,2\pi)$ and are mutually orthogonal. The squares of their L^2-norms are given by

$$\begin{cases} \displaystyle\int_0^{2\pi}\Phi_0^2(\varphi)\,d\varphi=\int_0^{2\pi}d\varphi=2\pi; \\ \displaystyle\int_0^{2\pi}\cos^2(k\varphi)\,d\varphi=\int_0^{2\pi}\sin^2(k\varphi)\,d\varphi=\pi, \quad k=1,2,3,\dots. \end{cases} \tag{17.19}$$

The Fourier method for the problem (17.16) in the annulus consists of finding the solution in the form of a series over the eigenfunctions of the problem (17.18):

$$u(\varphi,r)=\sum_{k=0}^{\infty}R_k(r)\cos k\varphi+\sum_{k=1}^{\infty}S_k(r)\sin k\varphi. \tag{17.20}$$

Substituting this series into equation (17.16), we get the following equations on the "radial" functions $R_k(r)$:

$$R_k''+\frac{1}{r}R_k'+\frac{1}{r^2}R_k(-k^2)=0, \qquad r_1<r<r_2, \quad k=0,1,2,\dots \tag{17.21}$$

and the same equations on S_k:

$$S_k''+\frac{1}{r}S_k'+\frac{1}{r^2}S_k(-k^2)=0, \qquad r_1<r<r_2, \quad k=0,1,2,\dots. \tag{17.22}$$

Let us solve the *radial* equations (17.21), (17.22). These are *the Euler equations*. Substituting $R_k=r^\lambda$ into (17.21), we get

$$\lambda(\lambda-1)r^{\lambda-2}+\lambda r^{\lambda-2}-k^2 r^{\lambda-2}=0,$$

and we get the characteristic equation $\lambda^2-k^2=0$, hence $\lambda=\pm k$. If $k\neq 0$, then the roots are simple, and the general solutions to (17.21) and (17.22) have the following form:

$$R_k(r)=A_k r^k+B_k r^{-k}, \qquad k=1,2,3,\dots; \tag{17.23}$$

$$S_k(r)=C_k r^k+D_k r^{-k}, \qquad k=1,2,3,\dots. \tag{17.24}$$

For $k=0$, the root of the equation $\lambda=0$ has multiplicity 2, hence

$$R_0(r) = A_0 + B_0 \ln r. \tag{17.25}$$

Substituting (17.23)–(17.25) into (17.20), we get the general solution of a homogeneous Laplace equation in the annulus:

$$u(\varphi, r) = A_0 + B_0 \ln r + \sum_{k=1}^{\infty} \left(A_k r^k + \frac{B_k}{r^k} \right) \cos k\varphi + \sum_{k=1}^{\infty} \left(C_k r^k + \frac{D_k}{r^k} \right) \sin k\varphi. \tag{17.26}$$

Remark 17.6. This is a general form of a harmonic function in the annulus.

The values of the constants in (17.26) are obtained from the boundary conditions (17.16):

$$\begin{cases} A_0 + B_0 \ln r_1 + \sum_{k=1}^{\infty} (A_k r_1^k + B_k r_1^{-k}) \cos k\varphi + \sum_{k=1}^{\infty} (C_k r_1^k + D_k r_1^{-k}) \sin k\varphi = f_1(\varphi), \\ A_0 + B_0 \ln r_2 + \sum_{k=1}^{\infty} (A_k r_2^k + B_k r_2^{-k}) \cos k\varphi + \sum_{k=1}^{\infty} (C_k r_2^k + D_k r_2^{-k}) \sin k\varphi = f_2(\varphi), \end{cases}$$
$$\tag{17.27}$$

where $0 \le \varphi \le 2\pi$. Taking into account the orthogonality of the eigenfunctions of the Sturm – Liouville problem (17.18) and the relations (17.19), we get

$$\begin{cases} A_0 + B_0 \ln r_1 = \dfrac{1}{2\pi} \displaystyle\int_0^{2\pi} f_1(\varphi) \, d\varphi, \\ A_0 + B_0 \ln r_2 = \dfrac{1}{2\pi} \displaystyle\int_0^{2\pi} f_2(\varphi) \, d\varphi, \end{cases} \tag{17.28}$$

and, similarly, for $k = 1, 2, 3, \ldots$,

$$\begin{cases} A_k r_1^k + B_k r_1^{-k} = \dfrac{1}{\pi} \displaystyle\int_0^{2\pi} f_1(\varphi) \cos k\varphi \, d\varphi, \\ A_k r_2^k + B_k r_2^{-k} = \dfrac{1}{\pi} \displaystyle\int_0^{2\pi} f_2(\varphi) \cos k\varphi \, d\varphi; \end{cases} \tag{17.29}$$

$$\begin{cases} C_k r_1^k + D_k r_1^{-k} = \dfrac{1}{\pi} \displaystyle\int_0^{2\pi} f_1(\varphi) \sin k\varphi \, d\varphi, \\ C_k r_2^k + D_k r_2^{-k} = \dfrac{1}{\pi} \displaystyle\int_0^{2\pi} f_2(\varphi) \sin k\varphi \, d\varphi. \end{cases} \tag{17.30}$$

We find A_0 and B_0 from the system (17.28) and A_k, B_k from (17.29). C_k and D_k are found from (17.30). The problem (17.15) is solved.

Problem 17.7. Prove that the solution (17.26) of the problem (17.15) is infinitely differentiable in the interior of the annulus.

Problem 17.8. Solve the Dirichlet problem in the annulus:

$$\begin{cases} \Delta u(x,y) = 0, & 4 < x^2 + y^2 < 9; \\ u|_{x^2+y^2=4} = x, & u|_{x^2+y^2=9} = y. \end{cases}$$

Solution. Here $r_1 = 2$, $r_2 = 3$, so that

$$f_1(\varphi) = 2\cos\varphi, \qquad f_2(\varphi) = 3\sin\varphi. \qquad (17.31)$$

Therefore, the right-hand sides in (17.28) are equal to zero and $A_0 = B_0 = 0$. Analogously, the right-hand sides of the systems (17.29) and (17.30) are equal to zero for all $k \neq 1$, thus

$$A_k = B_k = 0, \quad C_k = D_k = 0 \quad \text{for} \quad k \neq 1.$$

Hence, the series (17.26) contains only two terms:

$$u(\varphi, r) = (A_1 r + B_1 r^{-1})\cos\varphi + (C_1 r + D_1 r^{-1})\sin\varphi. \qquad (17.32)$$

The remaining coefficients are obtained from the systems of equations

$$\begin{cases} A_1 2 + B_1 \frac{1}{2} = 2, \\ A_1 3 + B_1 \frac{1}{3} = 0, \end{cases} \qquad \begin{cases} C_1 2 + D_1 \frac{1}{2} = 0, \\ C_1 3 + D_1 \frac{1}{3} = 3, \end{cases} \qquad (17.33)$$

which are derived directly from (17.31). Namely, (17.33) is obtained by substituting (17.31) into (17.27) and comparing the Fourier coefficients in both sides of the relations, instead of evaluating integrals in (17.29)–(17.30). From (17.33) we find

$$A_1 = -\frac{4}{5}, \qquad B_1 = \frac{36}{5}, \qquad C_1 = \frac{9}{5}, \qquad D_1 = -\frac{36}{5}. \qquad (17.34)$$

Answer. $u(\varphi, r) = \left(-\frac{4}{5}r + \frac{36}{5}r^{-1}\right)\cos\varphi + \left(\frac{9}{5}r - \frac{36}{5}r^{-1}\right)\sin\varphi.$

B. Now let us consider the Dirichlet problem in the disc of radius R:

$$\begin{cases} \triangle u(x,y) = 0, & x^2 + y^2 < R^2; \\ u|_{x^2+y^2=R^2} = f(\varphi), & 0 < \varphi < 2\pi. \end{cases} \qquad (17.35)$$

A solution of this problem also has the form (17.26), since the disc $x^2 + y^2 < R^2$ contains the (degenerate) annulus $0 < x^2 + y^2 < R^2$. But the disc also contains the point $(0,0)$, where the solution has to be finite:

$$|u(0,0)| < \infty. \qquad (17.36)$$

It can be shown [TS90] that (17.36) holds if and only if all the terms which have the singularity at $(0,0)$ of the form $\ln r$ and r^{-k} are absent from (17.26). This means that $B_0 = B_k = D_k = 0$, $k = 1, 2, 3, \dots$. Thus, (17.26) takes the form

$$u(x,y) = A_0 + \sum_{k=1}^{\infty} r^k (A_k \cos k\varphi + C_k \sin k\varphi). \qquad (17.37)$$

This is the analog of the Taylor series for a harmonic function in a disc. The coefficients of the series (17.37) are found from the boundary condition of the problem (17.35).

Problem 17.9. Solve the Dirichlet problem in the disc:

$$\begin{cases} \triangle u(x,y) = 0, & x^2 + y^2 < 4; \\ u|_{x^2+y^2=4} = x^2. \end{cases}$$

Solution. We are looking for the solution u in the form (17.37). The substitution of this series into the boundary condition gives:

$$A_0 + \sum_{k=1}^{\infty} 2^k (A_k \cos k\varphi + C_k \sin k\varphi) = 2 + 2\cos 2\varphi, \qquad (17.38)$$

since $x^2|_{r=2} = (2\cos\varphi)^2 = 4\cos^2\varphi = 2 + 2\cos 2\varphi$. Comparing the Fourier coefficients in the left- and right-hand sides of (17.38), we see that all A_k and C_k with $k \neq 0$ and $k \neq 2$ are equal to zero, and and the formula (17.37) yields the answer: $A_0 = 2$, $A_2 = 1/2$, $C_2 = 0$. The formula (17.37) takes the form

$$u = 2 + r^2 \frac{1}{2} \cos 2\varphi = 2 + \frac{r^2}{2}(\cos^2\varphi - \sin^2\varphi) = 2 + \frac{x^2 - y^2}{2}.$$

Problem 17.10. Solve the Dirichlet problem in the annulus:

$$\begin{cases} \triangle u(x,y) = x^2, & 9 < x^2 + y^2 < 16; \\ u|_{x^2+y^2=9} = 0, & u|_{x^2+y^2=16} = 0. \end{cases}$$

Hint. Both the solution that we are looking for and the right-hand side of the equation are to be decomposed into the series of the form (17.20). Equations on the radial functions R_k and S_k will be the nonhomogeneous Euler equations.

Problem 17.11. Solve the Neumann problem in the disc:

$$\begin{cases} \triangle u(x,y) = 0, & x^2 + y^2 < 9; \\ \frac{\partial u}{\partial n}\big|_{x^2+y^2=9} = y, \end{cases}$$

where $\frac{\partial}{\partial n}$ is the derivative normal to the boundary of the disc.

Hint. Solution is to be looked for in the form of the series (17.37); moreover, in the polar coordinates one has $\frac{\partial u}{\partial n} = \frac{\partial u}{\partial r}$.

Conclusion. The heat equation, the wave equation, and the Laplace equation possess different properties. As it follows from the results of Chapter 2, solutions of the homogeneous Laplace equation and the heat equation are smooth inside the regions where they are considered, even if the boundary values are discontinuous. At the same time, solutions of the homogeneous wave equation could be discontinuous if, for example, the initial data are discontinuous functions.

Chapter 3
Distributions and Green's functions

Laurent Schwartz introduced distributions in late 1940s. See [Sch66a, Sch66b]. "Weak derivatives" were introduced by S.L. Sobolev in 1935.

18 Motivation

Continuous functions $u(x) \in C(\mathbb{R})$ can be defined using the following three ways.
1. A continuous function can be uniquely specified by its values

$$\{u(x) \,:\, x \in \mathbb{R}\}. \tag{18.1}$$

2. It can be defined using its Fourier coefficients (if it is 2π-periodic):

$$u(x) = \sum_{k \in \mathbb{Z}} u_k e^{ikx}. \tag{18.2}$$

Here

$$u_k = \frac{1}{2\pi} \int_0^{2\pi} e^{-ikx} u(x) \, dx.$$

The sequence

$$\{u_k \,:\, k \in \mathbb{Z}\}$$

uniquely defines a continuous (periodic) function by the formula (18.2).
3. Let us introduce the space of the *test functions*. Let $C_0^\infty(\mathbb{R})$ be the space of smooth functions with the compact support; that is,

a. $\varphi(x) \in C^\infty(\mathbb{R})$;
b. $\varphi(x) \equiv 0$ for $|x| \geq A$, where $A \geq 0$ depends on φ (Fig. 18.1).

For any continuous function $u(x)$ define the scalar product with $\varphi \in C_0^\infty(\mathbb{R})$:

Alexander Komech and Andrew Komech, *Principles of Partial Differential Equations*,
Problem Books in Mathematics, DOI 10.2007/978-1-4419-1096-7_3,
© Springer Science + Business Media, LLC 2009

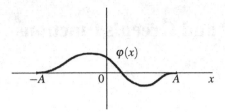

Fig. 18.1

$$\langle u, \varphi \rangle \equiv \int\limits_{-\infty}^{\infty} u(x)\varphi(x)\,dx. \tag{18.3}$$

This integral converges, since $\varphi(x) \equiv 0$ for $|x| \geq A$:

$$\langle u, \varphi \rangle = \int\limits_{-A}^{A} u(x)\varphi(x)\,dx.$$

For a particular continuous function $u(x)$, consider the set of values

$$\{\langle u, \varphi \rangle \; : \; \varphi \in C_0^{\infty}(\mathbb{R})\}. \tag{18.4}$$

Question 18.1. Is the function $u(x)$ uniquely defined by this set of values?

Answer. Yes. (Prove this!)

Question 18.2. Can the formula be written for restoring the continuous function $u(x)$ from the set of values (18.4)?

Answer. Yes:

$$u(x) = \lim_{\varepsilon \to 0} \frac{1}{\varepsilon} \int\limits_{-\infty}^{\infty} \varphi(\frac{x-y}{\varepsilon})u(y)\,dy = \lim_{\varepsilon \to 0}\langle \varphi_\varepsilon^x(y), u(y)\rangle. \tag{18.5}$$

Here $\varphi_\varepsilon^x(y) = \frac{1}{\varepsilon}\varphi(\frac{x-y}{\varepsilon}) \in C_0^{\infty}(\mathbb{R})$. The function $\varphi \in C_0^{\infty}(\mathbb{R})$ satisfies the following conditions (see Fig. 18.2):

$$\varphi(y) \equiv 0 \;\text{ for }\; |y| \geq 1; \qquad \int\limits_{-1}^{1} \varphi(y)\,dy = 1.$$

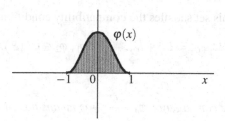

Fig. 18.2

Let us prove (18.5). We change the variable of integration to $z = \frac{x-y}{\varepsilon}$. Then (18.5) takes the form

$$u(x) = \lim_{\varepsilon \to 0} \int_{-1}^{1} \varphi(z) u(x - \varepsilon z) \, dz.$$

This formula follows from the continuity of u at the point x:

$$\lim_{\varepsilon \to 0} \int_{-1}^{1} \varphi(z) u(x - \varepsilon z) \, dz = \int_{-1}^{1} \varphi(z) \lim_{\varepsilon \to 0} u(x - \varepsilon z) \, dz = \int_{-1}^{1} \varphi(z) u(x) \, dz = u(x).$$

Question 18.3. What is the essential difference of the three ways of defining a function $u(x)$ which we described above?

Answer.
1. The set of numbers $\{u(x) : x \in \mathbb{R}\}$ can be more or less arbitrary: at any finite set of points $x_k \in \mathbb{R}$ the values $u(x_k)$ can be arbitrary.
2. The set of numbers $\{u_k : k \in \mathbb{Z}\}$ can be arbitrary, as long as $|u_k|$ decay for $|k| \to \infty$ fast enough so that, for example,

$$\sum_{k=-\infty}^{\infty} |u_k| < \infty.$$

3. The values $\{\langle u, \varphi \rangle : \varphi \in C_0^\infty(\mathbb{R})\}$ are not arbitrary: As could be seen from (18.3), if $l_\varphi = \langle u, \varphi \rangle$, then the numbers l_φ are connected by the algebraic relations

$$l_{\varphi_1 + \varphi_2} = \langle u, \varphi_1 + \varphi_2 \rangle = \langle u, \varphi_1 \rangle + \langle u, \varphi_2 \rangle = l_{\varphi_1} + l_{\varphi_2}, \qquad (18.6)$$

for all $\varphi_1, \varphi_2 \in C_0^\infty(\mathbb{R})$.

Conclusion. For the abstract set of numbers $\{l_\varphi : \varphi \in C_0^\infty(\mathbb{R})\}$ to correspond to some function $u(x) \in C(\mathbb{R})$ such that

$$l_\varphi = \langle u, \varphi \rangle, \qquad \forall \varphi \in C_0^\infty(\mathbb{R}), \tag{18.7}$$

it is necessary that this set satisfies the compatibility conditions (18.6):

$$l_{\varphi_1 + \varphi_2} = l_{\varphi_1} + l_{\varphi_2}, \qquad \forall \varphi_1, \varphi_2 \in C_0^\infty(\mathbb{R}). \tag{18.8}$$

Definition 18.4. *The convergence* $\varphi_n \xrightarrow[n \to \infty]{C_0^\infty} \varphi$ *means the following:*

a. $\varphi_n(x)$ *converges to* $\varphi(x)$ *uniformly in* $x \in \mathbb{R}$, *and the same is true for derivatives of any order:* $\forall k = 0, 1, 2, \ldots$

$$\varphi_n^{(k)}(x) \rightrightarrows \varphi^{(k)}(x), \quad x \in \mathbb{R}, \quad \text{as} \quad n \to \infty. \tag{18.9}$$

b. *All* φ_n *are supported inside* $[-A, A]$: *there is* $A > 0$ *such that for any* $n \in \mathbb{N}$

$$\varphi_n(x) \equiv 0 \quad \text{for} \quad |x| \geq A. \tag{18.10}$$

Question 18.5. Are the compatibility conditions (18.8) sufficient for the existence of a function $u(x) \in C(\mathbb{R})$ corresponding to the set $\{l_\varphi : \varphi \in C_0^\infty(\mathbb{R})\}$ in the sense of the identity (18.7)?

Answer. No, they are not. One also needs the continuity conditions:

$$l_{\varphi_n} \xrightarrow[n \to \infty]{} l_\varphi \quad \text{if} \quad \varphi_n \xrightarrow{C_0^\infty} \varphi \quad \text{for} \quad n \to \infty. \tag{18.11}$$

Indeed, under the conditions (18.9) and (18.10), given $u(x) \in C^1(\mathbb{R})$, then the convergence of $\langle u, \varphi_n \rangle$ to $\langle u, \varphi \rangle$ follows from interchanging the integration and taking the limit for uniformly convergent functions supported on a bounded interval:

$$\langle u, \varphi_n \rangle \equiv \int_{-A}^{A} u(x)\varphi_n(x)\,dx \xrightarrow[n \to \infty]{} \int_{-A}^{A} u(x)\varphi(x)\,dx = \langle u, \varphi \rangle,$$

since $u(x)\varphi_n(x) \rightrightarrows u(x)\varphi(x)$ as $x \in [-A, A]$, $n \to \infty$.

Thus, for the existence of a function $u(x) \in C(\mathbb{R})$ which corresponds to the set $\{l_\varphi : \varphi \in C_0^\infty(\mathbb{R})\}$ in the sense of (18.7), besides (18.8), the following condition is necessary:

$$l_{\varphi_n} \to l_\varphi \quad \text{if} \quad \varphi_n \xrightarrow{C_0^\infty} \varphi. \tag{18.12}$$

Question 18.6. Do the conditions (18.8) and (18.12) suffice for the existence of $u(x) \in C(\mathbb{R})$ corresponding to the representation (18.7)?

Answer. No.

Problem 18.7. Give the example of a set $\{l_\varphi \; : \; \varphi \in C_0^\infty(\mathbb{R})\}$ which satisfies the conditions (18.8) and (18.12), but such that there is no corresponding function $u(x) \in C(\mathbb{R})$ (see [Vla84]).

Answer.

$$l_\varphi = \varphi(0), \qquad \forall \varphi \in C_0^\infty(\mathbb{R}). \tag{18.13}$$

Conclusion. The set of values $\{l_\varphi \; : \; \varphi \in C_0^\infty(\mathbb{R})\}$ defines a function $u(x) \in C(\mathbb{R})$ satisfying the identity (18.7) uniquely only if such function $u(x)$ exists, although it may not exist for a particular $\{l_\varphi \; : \; \varphi \in C_0^\infty(\mathbb{R})\}$. The conditions (18.8), (18.12) are necessary for the existence of a continuous function $u(x)$, but not sufficient.

19 Distributions

Definition 19.1. *A distribution is a set* $l = \{l_\varphi \; : \; \varphi \in C_0^\infty(\mathbb{R})\}$ *that satisfies the conditions (18.8) and (18.12).*

For brevity, we denote

$$\mathscr{D} = \mathscr{D}(\mathbb{R}) = C_0^\infty(\mathbb{R}).$$

Remark 19.2. From the functional analysis point of view, the set $\{l_\varphi \; : \; \varphi \in C_0^\infty(\mathbb{R})\}$ satisfying the conditions (18.8), (18.12) (a distribution) is a continuous linear functional on $\mathscr{D}(\mathbb{R})$, that is, the element of the dual space $\mathscr{D}'(\mathbb{R})$:

$$l = \{l_\varphi \; : \; \varphi \in \mathscr{D}(\mathbb{R})\} \in \mathscr{D}'(\mathbb{R}); \qquad l(\varphi) \equiv l_\varphi, \quad \forall \varphi \in \mathscr{D}(\mathbb{R}).$$

Thus, $\mathscr{D}'(\mathbb{R})$ is the space of distributions.

Notation 19.3. *For a distribution* $\{l_\varphi \; : \; \varphi \in \mathscr{D}(\mathbb{R})\}$ *its value* l_φ *on a test function* $\varphi(x)$ *will be denoted by* $l(\varphi)$, $\langle l, \varphi \rangle$, *and also* $\langle l(x), \varphi(x) \rangle$, *and will be called the scalar product of the distribution l with the test function* φ:

$$l_\varphi = l(\varphi) = \langle l, \varphi \rangle = \langle l(x), \varphi(x) \rangle. \tag{19.1}$$

Let us point out that $l(x)$ is not the value of the function l at the point x, but merely a symbolic notation.

Example 19.4. Distribution (18.13) is called the Dirac δ-function:

$$\delta_\varphi = \delta(\varphi) = \langle \delta, \varphi \rangle = \varphi(0), \qquad \forall \varphi \in \mathscr{D}(\mathbb{R}). \tag{19.2}$$

Remark 19.5. The formula (18.3) assigns a distribution to each continuous function $u(x) \in C(\mathbb{R})$:

$$u(x) \mapsto \{\langle u, \varphi \rangle \; : \; \varphi \in \mathscr{D}(\mathbb{R})\}.$$

According to (18.5), this mapping is injective:

$$C(\mathbb{R}) \subset \mathscr{D}'(\mathbb{R}).$$

But not every distribution could be represented by (18.3) using some continuous function (for example, $\delta(x)$ could not be represented in this way).

Let us consider examples of distributions.

a. For $k \in \mathbb{N}$, set

$$\langle d_k, \varphi \rangle = \varphi^{(k)}(0), \qquad \forall \varphi \in \mathscr{D}(\mathbb{R}).$$

Check that $d_k(x) \in \mathscr{D}'(\mathbb{R})$ (that is, check the conditions (18.8) and (18.11)).

b. The Heaviside function (see Fig. 19.1):

$$\Theta(x) = \begin{cases} 1, & x > 0; \\ 0, & x < 0; \end{cases} \tag{19.3}$$

$$\langle \Theta, \varphi \rangle \equiv \int_{-\infty}^{\infty} \Theta(x) \varphi(x)\, dx = \int_{0}^{\infty} \varphi(x)\, dx. \tag{19.4}$$

Check that $\Theta \in \mathscr{D}'(\mathbb{R})$.

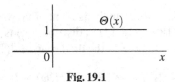

Fig. 19.1

20 Operations on distributions

Addition of distributions

Let us first notice that for continuous functions $u_1(x)$, $u_2(x)$ and their sum $u_1(x) + u_2(x)$ we have, due to (18.10),

$$\langle u_1 + u_2, \varphi \rangle = \langle u_1, \varphi \rangle + \langle u_2, \varphi \rangle, \qquad \forall \varphi \in \mathscr{D}(\mathbb{R}). \tag{20.1}$$

Definition 20.1. *For $l, m \in \mathscr{D}'(\mathbb{R})$ we set*

$$(l + m)_\varphi = l_\varphi + m_\varphi, \qquad \forall \varphi \in \mathscr{D}(\mathbb{R}). \tag{20.2}$$

Remark 20.2. Under such a definition, the addition of continuous functions $u_1(x)$, $u_2(x)$ coincides with the addition of the corresponding distributions. This is seen from (20.1) and (20.2).

Problem 20.3. Check that $l + m \in \mathscr{D}'(\mathbb{R})$.

Multiplication by a scalar

Let us notice that for any $u \in C(\mathbb{R})$, $\alpha \in \mathbb{R}$, and $\varphi \in \mathscr{D}(\mathbb{R})$, we have $\langle \alpha u, \varphi \rangle = \alpha \langle u, \varphi \rangle$.

Definition 20.4. *For $l \in \mathscr{D}'(\mathbb{R})$ and $\alpha \in \mathbb{R}$ we define*

$$\langle \alpha l, \varphi \rangle = \alpha \langle l, \varphi \rangle, \qquad \forall \varphi \in \mathscr{D}(\mathbb{R}).$$

Multiplication by a smooth function

Let $g \in C^\infty(\mathbb{R})$. If $u \in C(\mathbb{R})$, then, as seen from (18.3),

$$\langle gu, \varphi \rangle = \langle u, g\varphi \rangle, \qquad \forall \varphi \in \mathscr{D}(\mathbb{R}).$$

Definition 20.5. *For $l \in \mathscr{D}'(\mathbb{R})$, we set*

$$\langle gl, \varphi \rangle = \langle l, g\varphi \rangle, \qquad \forall \varphi \in \mathscr{D}(\mathbb{R}). \tag{20.3}$$

Remark 20.6. The right-hand side of (20.3) makes sense, since

$$g(x)\varphi(x) \in \mathscr{D}(\mathbb{R}). \tag{20.4}$$

Problem 20.7.

a. Verify that if $g \in C^\infty(\mathbb{R})$ and $g \in \mathscr{D}'(\mathbb{R})$, then $gl \in \mathscr{D}'(\mathbb{R})$.

b. For the Dirac δ-function, compute $x\delta$.

c. Prove that

$$g(x)\delta = g(0)\delta. \tag{20.5}$$

Shift of distributions

For $u(x) \in C(\mathbb{R})$ and $a \in (\mathbb{R})$,

$$\int_{-\infty}^{\infty} u(x-a)\varphi(x)\,dx = \int_{-\infty}^{\infty} u(y)\varphi(y+a)\,dy, \qquad \forall \varphi \in \mathscr{D}(\mathbb{R}).$$

Definition 20.8. *Let* $a \in \mathbb{R}$. *For* $l \in \mathscr{D}'(\mathbb{R})$, *we define its shift to the right by* a, *denoted* $l(x - a) \in \mathscr{D}'(\mathbb{R})$, *by*

$$\langle l(x - a), \varphi(x) \rangle = \langle l(y), \varphi(y + a) \rangle.$$

Remark 20.9. The notations $l(x)$ and $l(x - a)$ do not make the sense as functions (unless $l \in \mathscr{D}'(\mathbb{R})$ corresponds to a function defined pointwise). Yet, these notations reflect the fact that $l(x - a)$ is the shift of the distribution l by $a \in \mathbb{R}$. Such notations are common and useful in the analytical manipulations. For example,

$$\langle \delta(x - a), \varphi(x) \rangle = \langle \delta(y), \varphi(y + a) \rangle = \varphi(a). \qquad (20.6)$$

Change of scale (dilation) in the argument of distributions

For $u(x) \in C(\mathbb{R})$ and $k \neq 0$, $\displaystyle\int_{-\infty}^{+\infty} u(kx)\varphi(x)\,dx = \frac{1}{|k|} \int_{-\infty}^{+\infty} u(y)\varphi\left(\frac{y}{k}\right) dy.$

Definition 20.10. *For* $f(x) \in \mathscr{D}(\mathbb{R})$ *we set for* $k \neq 0$

$$\langle f(kx), \varphi(x) \rangle = \frac{1}{|k|} \left\langle f(y), \varphi\left(\frac{y}{k}\right) \right\rangle.$$

Problem 20.11. Prove that

$$\delta(kx) = \frac{1}{|k|}\delta(x), \qquad k \neq 0.$$

In particular, prove that δ is even:

$$\delta(-x) = \delta(x).$$

Remark 20.12. From the definition (20.6) we get:

$$\langle \delta(y - x), \varphi(y) \rangle = \varphi(x).$$

This means that $\delta(y - x)$ is the integral kernel of the unit operator $I\varphi = \varphi$. In the linear algebra the matrix of the unit operator is the Kronecker δ-symbol, δ_{ij}. It is due to this analogue that Dirac called functional (19.2) the δ-function.

Problem 20.13. Write the formula of a general change of variable $x = g(y)$ in a distribution, where $g : \mathbb{R} \to \mathbb{R}$ is a smooth diffeomorphism.

Convergence of distributions

Definition 20.14. *Distributions $u_n(x) \in \mathscr{D}'(\mathbb{R})$ converge (weakly) to $u(x) \in \mathscr{D}'(\mathbb{R})$ as $n \to \infty$ if for $\forall \varphi(x) \in C_0^\infty(\mathbb{R}) \equiv \mathscr{D}(\mathbb{R})$ one has*

$$\langle u_n, \varphi \rangle \to \langle u, \varphi \rangle \quad as \quad n \to \infty. \tag{20.7}$$

There is the following notation: $u_n(x) \xrightarrow{\mathscr{D}'(\mathbb{R})} u(x)$ *as $n \to \infty$.*

Examples of convergent series of distributions:

a. If $u_n \in C(\mathbb{R})$ and $u_n(x) \rightrightarrows u(x)$ as $n \to \infty$, then $u_n \xrightarrow{\mathscr{D}'(\mathbb{R})} u$ as $n \to \infty$. (Prove this!)

b. If $u_n(x) \in L^2(\mathbb{R})$ and $u_n \to u$ in $L^2(\mathbb{R})$ as $n \to \infty$, then $u_n \xrightarrow{\mathscr{D}'(\mathbb{R})} u$. (Prove this!)

c. $\sin kx \xrightarrow{\mathscr{D}'(\mathbb{R})} 0$ as $k \to \infty$. Indeed, integrating by parts, we obtain:

$$\langle \sin kx, \varphi(x) \rangle = \int \frac{\cos kx}{k} \varphi'(x)\, dx \longrightarrow 0 \quad as \quad k \to \infty.$$

d. Analogously, $k^2 \sin kx \xrightarrow{\mathscr{D}'(\mathbb{R})} 0$ as $k \to \infty$.

Remark 20.15. Sequences of functions $\sin kx$ and $k^2 \sin kx$ do not converge neither in the space $C(\mathbb{R})$, nor in the space $L^2(\mathbb{R})$, but they do converge in $\mathscr{D}'(\mathbb{R})$.

e. "δ-like" sequences. Consider the Steklov step-functions

$$u_n(x) = \begin{cases} n, & x \in [0, \frac{1}{n}]; \\ 0, & x \notin [0, \frac{1}{n}], \quad n \in \mathbb{N}. \end{cases} \tag{20.8}$$

Obviously, $\int\limits_{-\infty}^{+\infty} u_n(x)\, dx = 1$, $\forall n = 1, 2, 3, \ldots$.

Problem 20.16. Prove that

$$u_n(x) \xrightarrow{\mathscr{D}'(\mathbb{R})} \delta(x) \quad as \quad n \to \infty. \tag{20.9}$$

Hint. Apply the mean value theorem to the integral $\int_{-\infty}^{+\infty} u_n(x)\varphi(x)\, dx$.

Analogously, the Gauss distributions converge weakly to a δ-function (prove this!):

$$\frac{e^{-\frac{x^2}{2\sigma}}}{\sqrt{2\pi\sigma}} \xrightarrow{\mathscr{D}'(\mathbb{R})} \delta(x) \quad as \quad \sigma \to 0+.$$

Differentiation of distributions

Let $\varphi \in \mathscr{D}(\mathbb{R})$, $\varphi(x) \equiv 0$ for $|x| \geq A$. For $u(x) \in C^1(\mathbb{R})$, integrating by parts, we get:

$$\int_{-\infty}^{+\infty} u'(x)\varphi(x)\,dx = u\varphi\Big|_{-A}^{A} - \int_{-A}^{A} u(x)\varphi'(x)\,dx = -\int_{-\infty}^{+\infty} u(x)\varphi'(x)\,dx.$$

The boundary term is equal to zero since $\varphi(A) = \varphi(-A) = 0$.

Definition 20.17. *For $u \in \mathscr{D}'(\mathbb{R})$ we set*

$$\langle u', \varphi \rangle = -\langle u, \varphi' \rangle, \qquad \forall \varphi \in \mathscr{D}(\mathbb{R}). \tag{20.10}$$

Problem 20.18. Prove that $u' \in \mathscr{D}'(\mathbb{R})$.

Thus,

> Any distribution has a derivative which is also a distribution, and hence derivatives of all orders!

Let us consider examples of differentiation of distributions.

a. $(\sin x)' = \cos x$.
b. Let us find Θ' (see (19.3), (19.4)). According to the definition (20.10),

$$\langle \Theta', \varphi \rangle = -\langle \Theta, \varphi' \rangle = \int_{0}^{+\infty} \varphi'(x)\,dx = -\varphi(x)\Big|_{0}^{+\infty} = \varphi(0) = \langle \delta, \varphi \rangle.$$

From here, we see that
$$\Theta'(x) = \delta(x). \tag{20.11}$$

Continuity of the differentiation in the sense of distributions

Lemma 20.1. The operator $\frac{d}{dx} : \mathscr{D}'(\mathbb{R}) \to \mathscr{D}'(\mathbb{R})$ is continuous.

Proof. Let $u_n \xrightarrow{\mathscr{D}'(\mathbb{R})} u$. Then for $\forall \varphi \in \mathscr{D}(\mathbb{R})$,

$$\langle u_n', \varphi \rangle \equiv -\langle u_n, \varphi' \rangle \xrightarrow[n\to\infty]{} -\langle u, \varphi' \rangle \equiv -\langle u', \varphi \rangle.$$

Consequently, $u_n' \xrightarrow{\mathscr{D}'(\mathbb{R})} u'$ according to the definition (20.7).

Problem 20.19. Prove that for for any $u \in \mathscr{D}'(\mathbb{R})$,

$$\frac{u(x+\varepsilon) - u(x)}{\varepsilon} \xrightarrow{\mathscr{D}'(\mathbb{R})} u' \quad \text{as} \quad \varepsilon \to 0.$$

21 Differentiation of jumps and the product rule

Differentiation of jumps

Lemma 21.1. Let $u(x) \in C^1$ for $x < a$ and for $x > a$, while at the point $x = a$ it has the jump discontinuity; that is, the one-sided limits $u(a \pm 0) := \lim_{x \to a \pm 0} a(x)$ exist and are different (for simplicity, we assume that $u'(a \pm 0)$ also exist). See Fig. 21.1.

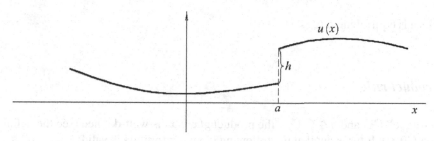

Fig. 21.1

Then the following formula is valid:

$$u'(x) = \{u'(x)\} + h \cdot \delta(x - a), \qquad h := u(a+0) - u(a-0). \tag{21.1}$$

The function $u'(x)$ in the left-hand side of (21.1) is a generalized derivative of the distribution $u(x)$, while $\{u'(x)\}$ in the right-hand side is a function continuous for $x \neq a$, which is equal to the derivative of the function $u(x)$ at the points where this derivative exists. The distribution given for $\{u'(x)\}$ by the formula (18.3) is called the *regular part* of the generalized derivative $u'(x)$.

Example 21.1. For $u(x) = \Theta(x)$, we have: $a = 0$, $\{\Theta'(x)\} \equiv 0$, since $\Theta'(x) = 0$ for $x \neq 0$, $h = \Theta(0+) - \Theta(0-) = 1$. Therefore, in agreement with (20.11), the formula (21.1) gives

$$\Theta'(x) = \delta(x). \tag{21.2}$$

Problem 21.2. Compute $|x|''$.

Solution. According to the formula (21.1),

$$|x|' = \{|x|'\} + 0 \cdot \delta(x) = \operatorname{sgn} x \equiv \begin{cases} 1, & x > 0; \\ -1, & x < 0. \end{cases}$$

Again, using the same formula,

$$|x|'' = (\operatorname{sgn} x)' = \{\operatorname{sgn}' x\} + 2\delta(x) = 2\delta(x). \tag{21.3}$$

Problem 21.3. Prove the formula (21.1).

Solution. For $\varphi \in C_0^\infty(\mathbb{R})$,

$$\langle u', \varphi \rangle = -\langle u, \varphi' \rangle = -\int_{-\infty}^{a} u(x)\varphi'(x)\,dx - \int_{a}^{+\infty} u(x)\varphi'(x)\,dx$$

$$= -u\varphi\Big|_{-\infty}^{a-0} - u\varphi\Big|_{a+0}^{\infty} + \int_{x\neq a} u'(x)\varphi(x)\,dx = u(a)(\varphi(a+0) - \varphi(a-0)) + \langle\{u'\}, \varphi\rangle,$$

which is equivalent to (21.1).

Product rule

For $g \in C^\infty(\mathbb{R})$ and $u \in \mathscr{D}'(\mathbb{R})$, the product $g(x)u(x)$ is well-defined (see the definition (20.3)). It turns out that the following common formula is valid:

$$(gu)' = g'u + gu'. \tag{21.4}$$

Problem 21.4. Prove the formula (21.4).

Problem 21.5. Using the formula (21.4), compute the following:

$$\left(\frac{d}{dx} + \lambda\right)(\Theta(x)e^{-\lambda x}). \tag{21.5}$$

Solution. According to formulas (21.4) and (20.5),

$$\frac{d}{dx}(e^{-\lambda x}\Theta(x)) = -\lambda e^{-\lambda x}\Theta(x) + e^{-\lambda x}\Theta'(x) = -\lambda e^{-\lambda x}\Theta(x) + \delta(x).$$

Adding $\lambda\Theta(x)e^{-\lambda x}$ to both sides, we see that (21.5) is equal to $\delta(x)$:

$$\left(\frac{d}{dx} + \lambda\right)(\Theta(x)e^{-\lambda x}) = \delta(x). \tag{21.6}$$

Problem 21.6. For $\omega \neq 0$, compute $\left(\frac{d^2}{dx^2} + \omega^2\right)\left(\Theta(x)\frac{\sin \omega x}{\omega}\right) = \ldots$.

Solution. According to formulae (21.4) and (20.5),

$$\frac{d}{dx}\left(\Theta(x)\frac{\sin \omega x}{\omega}\right) = \Theta(x)\cos \omega x + \Theta'(x)\frac{\sin \omega x}{\omega} = \Theta(x)\cos \omega x.$$

Using the same formulae, we get

$$\frac{d^2}{dx^2}\left(\Theta(x)\frac{\sin\omega x}{\omega}\right) = \frac{d}{dx}(\Theta(x)\cos\omega x)$$
$$= -\omega\Theta(x)\sin\omega x + \Theta'(x)\cos\omega x = -\omega\Theta(x)\sin\omega x + \delta(x).$$

Substituting $\omega^2\Theta(x)\frac{\sin\omega x}{\omega}$, we get $\delta(x)$:

$$\left(\frac{d^2}{dx^2} + \omega^2\right)\left(\Theta(x)\frac{\sin\omega x}{\omega}\right) = \delta(x). \qquad (21.7)$$

Problem 21.7. Prove (21.6) and (21.7) using formula (21.1) instead of (21.4).

Solution. Let us prove (21.6). We plot $\Theta(x)e^{-\lambda x}$ (Fig. 21.2).

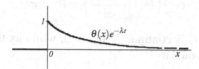

Fig. 21.2

According to the formula (21.1) (with $a = 0$ and $h = 1$),

$$\frac{d}{dx}(\Theta(x)e^{-\lambda x}) = \Theta(x)(-\lambda)e^{-\lambda x} + \delta(x).$$

Adding $\lambda\Theta(x)e^{-\lambda x}$, we get (21.6).

Let us prove (21.7). The plot of $\Theta(x)\frac{\sin\omega x}{\omega}$ is on Fig. 21.3.

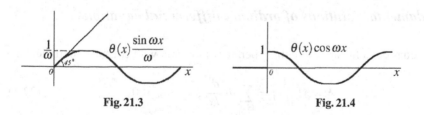

Fig. 21.3   **Fig. 21.4**

According to the formula (21.1) (with $a = 0$ and $h = 0$),

$$\frac{d}{dx}\left(\Theta(x)\frac{\sin\omega x}{\omega}\right) = \Theta(x)\cos\omega x.$$

The graph of $\Theta(x)\cos\omega x$ is plotted on Fig. 21.4.
We use the formula (21.1) (with $a = 0$ and $h = 1$):

$$\frac{d^2}{dx^2}\left(\Theta(x)\frac{\sin\omega x}{\omega}\right) = \frac{d}{dx}\left(\Theta(x)\cos\omega x\right) = -\omega\Theta(x)\sin\omega x + \delta(x).$$

Adding $\omega^2\Theta(x)\frac{\sin\omega x}{\omega}$, we get (21.7).

Remark 21.8. We have the equality (21.6) since the function $\Theta(x)e^{-\lambda x}$ for $x \neq 0$ satisfies the homogeneous equation

$$\left(\frac{d}{dx}+\lambda\right)\left(\Theta(x)e^{-\lambda x}\right) = 0 \quad \text{for} \quad x \neq 0, \tag{21.8}$$

while its jump is $h = 1$. Analogously, we have the equality (21.7) since the function $y(x) = \Theta(x)\frac{\sin\omega x}{\omega}$ for $x \neq 0$ satisfies the homogeneous equation

$$\left(\frac{d^2}{dx^2}+\omega^2\right)\left(\Theta(x)\frac{\sin\omega x}{\omega}\right) = 0 \quad \text{for} \quad x \neq 0. \tag{21.9}$$

Besides, the function $y(x)$ is continuous at $x = 0$, while its first derivative $y'(x) = \Theta(x)\cos\omega x$ has a jump equal to 1:

$$\begin{cases} y(0-) = y(0+), \\ y'(0+) = y'(0-) + 1. \end{cases} \tag{21.10}$$

Thus, the regular parts in (21.6) and (21.7) cancel out due to equations (21.8) and (21.9), respectively.

22 Fundamental solutions of ordinary differential equations

Fundamental solutions of ordinary differential equations

Let us consider a linear differential operator of order m with constant coefficients:

$$A = A\left(\frac{d}{dx}\right) = \sum_{k=0}^{m} a_k \frac{d^k}{dx^k}, \qquad a_m \neq 0. \tag{22.1}$$

Using the chain rule, we get $\frac{d^k}{dx^k}u(x-y) = u^{(k)}(x-y)$, $x \in \mathbb{R}$. Therefore,

$$A\left(\frac{d}{dx}\right)u(x-y) = (Au)(x-y), \qquad x \in \mathbb{R}. \tag{22.2}$$

Definition 22.1. *The fundamental solution of the operator A is a distribution $\mathscr{E}(x) \in \mathscr{D}'(\mathbb{R})$ such that*

$$A\left(\frac{d}{dx}\right)\mathscr{E}(x) = \delta(x), \qquad x \in \mathbb{R}, \tag{22.3}$$

where the derivatives are understood in the sense of distributions.

Remark 22.2. As it follows from (22.2), $A\left(\frac{d}{dx}\right)\mathscr{E}(x-y)=\delta(x-y)$, $x\in\mathbb{R}$.

Examples.

a. For $A=\frac{d}{dx}$, the fundamental solution is $\mathscr{E}(x)=\Theta(x)$.   See (21.2).
b. For $A=\frac{d^2}{dx^2}$, $\mathscr{E}(x)=\frac{1}{2}|x|$.   See (21.3).
c. For $A=\frac{d}{dx}+\lambda$, $\mathscr{E}(x)=\Theta(x)e^{-\lambda x}$.   See (21.6).
d. For $A=\frac{d^2}{dx^2}+\omega^2$, $\mathscr{E}(x)=\Theta(x)\frac{\sin\omega x}{\omega}$.   See (21.7).

Let us point out that for a fixed operator A there could be infinitely many fundamental solutions.

Question 22.3. Why does one need fundamental solutions?

Answer. To solve nonhomogeneous equations

$$A\left(\frac{d}{dx}\right)u(x)=f(x),\qquad x\in\mathbb{R}.$$

A particular solution could be found using the formula

$$u(x)=\int_{-\infty}^{+\infty}\mathscr{E}(x-y)f(y)\,dy\equiv(\mathscr{E}*f)(x)=\int_{-\infty}^{+\infty}\mathscr{E}(y)f(x-y)\,dy,\qquad(22.4)$$

if $f(x)=0$ for $|x|\ge$ const and $f(x)\in C(\mathbb{R})$. The operation $*$ in (22.4) is called a convolution of $\mathscr{E}$ with f.

Let us prove this for the case $f(x)\in C^m(\mathbb{R})$: For the function (22.4) we get from (22.3), for $x\in\mathbb{R}$:

$$A\left(\frac{d}{dx}\right)u(x)=\int_{-\infty}^{+\infty}\mathscr{E}(y)A\left(\frac{d}{dx}\right)f(x-y)\,dy=\left\langle\mathscr{E}(y),A\left(-\frac{d}{dy}\right)f(x-y)\right\rangle$$

$$=\left\langle A\left(\frac{d}{dy}\right)\mathscr{E}(y),f(x-y)\right\rangle=\langle\delta(y),f(x-y)\rangle=f(x).$$

Examples.

a. For the equation $\frac{d}{dx}u(x)=f(x)$, $x\in\mathbb{R}$, the formula (22.4) gives a particular solution

$$u(x)=\int_{-\infty}^{+\infty}\Theta(x-y)f(y)\,dy=\int_{-\infty}^{x}f(y)\,dy,\qquad x\in\mathbb{R}.$$

b. For the equation

$$\frac{d^2}{dx^2}u(x)=f(x),\qquad x\in\mathbb{R}$$

the formula (22.4) gives a particular solution

$$u(x) = \int_{-\infty}^{+\infty} \frac{1}{2}|x-y|f(y)\,dy, \qquad x \in \mathbb{R},$$

which is analogous to the Cauchy formula for repeated integration,

$$u(x) = \int_{a}^{x} (x-y)f(y)\,dy.$$

Construction of fundamental solutions for arbitrary operators

For an operator A of the form (22.1), let $u_0(x)$ be a solution to the Cauchy problem

$$\begin{cases} A\left(\dfrac{d}{dx}\right) u_0(x) = 0, & x > 0; \\ u_0(0) = 0, \quad u_0'(0) = 0, \quad \dots, \quad u_0^{(m-2)}(0) = 0; \\ u_0^{(m-1)}(0) = \dfrac{1}{a_m}. \end{cases} \tag{22.5}$$

Then the function

$$\mathscr{E} = \begin{cases} u_0(x), & x > 0; \\ 0, & x < 0 \end{cases} \tag{22.6}$$

is the fundamental solution of the operator A.

Problem 22.4. Prove (22.3) for the function (22.6) using the formula (21.1).

Problem 22.5. Solve the equation

$$3u''(x) - u'(x) = \delta(x), \quad x \in \mathbb{R}.$$

Solution. The characteristic equation $3\lambda^2 - \lambda = 0$ gives $\lambda_1 = 0$ and $\lambda_2 = \frac{1}{3}$, hence

$$u_0(x) = c_1 + c_2 e^{\frac{x}{3}}.$$

The initial conditions in (22.5) give $c_1 + c_2 = 0$, $\frac{c_2}{3} = \frac{1}{3}$, hence $c_1 = -1$, $c_2 = 1$.
Answer. $u(x) = \Theta(x)(e^{\frac{x}{3}} - 1)$.

Problem 22.6. Find a particular solution to the equation

$$u''(x) - 3u'(x) + 2u(x) = f(x), \quad x \in \mathbb{R}$$

where $f(x) \in C(\mathbb{R})$, $f(x) = 0$ for $|x| \geq$ const.

Solution. Let us find the fundamental solution:

$$\mathscr{E}''(x) - 3\mathscr{E}'(x) + 2\mathscr{E}(x) = \delta(x).$$

The roots of the characteristic equation $\lambda^2 - 2\lambda + 2 = 0$ are $\lambda_1 = 1$, $\lambda_2 = 2$, hence

$$\mathscr{E}(x) = \Theta(x)(c_1 e^x + c_2 e^{2x}).$$

The initial conditions in (22.5) take the form $c_1 + c_2 = 0$, $c_1 + 2c_2 = 1$; hence, $c_1 = -1$, $c_2 = 1$.

Answer. According to the formula (22.4),

$$u(x) = \mathscr{E} * f(x) = \int_{-\infty}^{+\infty} \Theta(x-y)(e^{2(x-y)} - e^{x-y})f(y)\,dy = \int_{-\infty}^{x}(e^{2(x-y)} - e^{x-y})f(y)\,dy.$$

23 Green's function on an interval

Definition of Green's function

Let us find the solution to the following boundary value problem (where $\omega \neq 0$):

$$\begin{cases} u''(x) - \omega^2 u(x) = f(x), & 0 < x < l; \\ u(0) = u(l) = 0. \end{cases} \tag{23.1}$$

Green's function of this boundary value problem is a function $G(x,y)$ on $[0,l] \times [0,l]$, smooth for $x \neq y$ and satisfying the equations

$$\begin{cases} \left(\frac{d^2}{dx^2} - \omega^2\right)G(x,y) = \delta(x-y), & 0 < x < l; \\ G(0,y) = G(l,y) = 0. \end{cases} \tag{23.2}$$

Here y plays the role of a parameter, $y \in (0,l)$. We can say that Green's function is the fundamental solution that satisfies the boundary conditions.

Having Green's function, one can find the solution to the boundary value problem (23.1) using the formula

$$u(x) = \int_0^l G(x,y)f(y)\,dy. \tag{23.3}$$

Indeed, the boundary conditions (23.1) follow from the boundary conditions (23.2):
At $x = 0$,

$$u(0) = \int_0^l G(0,l)f(y)\,dy = 0$$

and similarly at $x = l$. Equation (23.1) can be checked formally:

$$\left(\frac{d^2}{dx^2} - \omega^2\right)u(x) = \int_0^l \left(\frac{d^2}{dx^2} - \omega^2\right)G(x,y)f(y)\,dy = \int_0^l \delta(x-y)f(y)\,dy = f(x).$$

Remark 23.1. The formula (23.3) means that Green's function $G(x,y)$ is the integral kernel of the operator G which is the inverse to the operator $A = \frac{d^2}{dx^2} - \omega^2$ of the boundary value problem (23.1):

$$A = \frac{d^2}{dx^2} - \omega^2: \quad C_0^2[0,l] \longrightarrow C[0,l]. \tag{23.4}$$

Here $C_0^2[0,l]$ is the space of functions $u(x) \in C^2[0,l]$ which satisfy the boundary conditions $u(0) = u(l) = 0$.

Remark 23.2. Operator (23.4) is symmetric, as was shown in (11.6). Hence, the operator $G = A^{-1}$ is also symmetric. It is here that the important symmetry property of Green's function is coming from:

$$G(x,y) = G(y,x), \qquad \forall x, y \in [0,l].$$

Construction of Green's function

The differential equation (23.2) is homogeneous for $x \neq y$, since $\delta(x-y) = 0$ for $x - y \neq 0$. Therefore, analogously to (21.9),

$$\left(\frac{d^2}{dx^2} - \omega^2\right)G(x,y) = 0 \quad \text{for} \quad x \neq y. \tag{23.5}$$

Therefore,

$$G(x,y) = \begin{cases} Ae^{\omega x} + Be^{-\omega x}, & x < y; \\ Ce^{\omega x} + De^{-\omega x}, & x > y. \end{cases} \tag{23.6}$$

For determining the constants A, B, C, and D, we have two boundary conditions in (23.2) and two jump conditions at $x = y$ (see Fig. 23.1), which are similar to (21.10):

$$\begin{cases} G(y-0,y) = G(y+0,y), \\ G'_x(y+0,y) = G'_x(y-0,y) + 1. \end{cases} \tag{23.7}$$

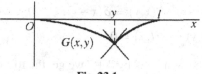

Fig. 23.1

These four equations determine A, B, C, and D uniquely.

Problem 23.3. Derive (23.2) from (23.5) and (23.7).

Hint. Apply the formula (21.1) (twice) for computing $\frac{d^2}{dx^2}G(x,y)$.

We point out that one could automatically take into account the boundary conditions (23.2) looking for Green's function in the form

$$G(x,y) = \begin{cases} A\sinh\omega x, & x < y; \\ B\sinh\omega(x-l), & x > y. \end{cases} \tag{23.8}$$

Then (23.6) is satisfied. It remains to take into account the jump conditions (23.7):

$$A\sinh\omega y = B\sinh\omega(y-l), \qquad B\omega\cosh\omega(y-l) = A\omega\cosh\omega y + 1.$$

Solving these two equations, we find

$$A = \frac{\sinh\omega(y-l)}{\omega\sinh\omega l}, \qquad B = \frac{\sinh\omega y}{\omega\sinh\omega l}. \tag{23.9}$$

Finally, from (23.8) we find Green's function for the problem (23.1):

$$G(x,y) = \begin{cases} \dfrac{\sinh\omega(y-l)\sinh\omega x}{\omega\sinh\omega l}, & x < y; \\[3mm] \dfrac{\sinh\omega y\sinh\omega(x-l)}{\omega\sinh\omega l}, & x > y. \end{cases} \tag{23.10}$$

Substituting into (23.3), we find the solution of the boundary value problem (23.1):

$$u(x) = \int_0^x \frac{\sinh\omega y\sinh\omega(x-l)}{\omega\sinh\omega l}f(y)\,dy + \int_x^l \frac{\sinh\omega(y-l)\sinh\omega x}{\omega\sinh\omega l}f(y)\,dy.$$

Note that Green's function (23.10) is symmetric, in agreement with Remark 23.2.

Problem 23.4. Let $\omega \neq 0$. Find the solution to the boundary value problem

$$\begin{cases} u''(x) + \omega^2 u(x) = f(x), & 0 < x < l; \\ u(0) = u(l) = 0. \end{cases} \tag{23.11}$$

Solution. We find Green's function $G(x,y)$ for $y \in [0,l]$:

$$\begin{cases} (\frac{d^2}{dx^2}+\omega^2)G(x,y)=\delta(x-y), & 0<x<l; \\ G(0,y)=G(l,y)=0. \end{cases}$$

Substituting in (23.5) the sign " $-$ " by " $+$ " we get the following (Cf. (23.8)):

$$G(x,y) = \begin{cases} A\sin\omega x, & x<y; \\ B\sin\omega(x-l), & x>y. \end{cases}$$

Substituting G into the jump conditions (23.7), we see that for $\sin\omega l \neq 0$,

$$A=\frac{\sin\omega(y-l)}{\omega\sin\omega l}, \qquad B=\frac{\sin\omega y}{\omega\sin\omega l}.$$

From here, similarly to (23.10),

$$G(x,y) = \begin{cases} \dfrac{\sin\omega(y-l)\sin\omega x}{\omega\sin\omega l}, & x<y; \\ \dfrac{\sin\omega y\sin\omega(x-l)}{\omega\sin\omega l}, & x>y. \end{cases} \tag{23.12}$$

Finally, we write down the solution to the problem (23.11):

$$u(x)=\int_0^x \frac{\sin\omega y\sin\omega(x-l)}{\omega\sin\omega l}f(y)\,dy+\int_x^l \frac{\sin\omega(y-l)\sin\omega x}{\omega\sin\omega l}f(y)\,dy. \tag{23.13}$$

Let us point out that Green's function (23.12) is also symmetric.

Problem 23.5. Find the solution to

$$\begin{cases} u''(x)-\omega^2 u(x)=f(x), & 0<x<l; \\ u(x)=u'(l)=0. \end{cases}$$

Problem 23.6. Find the solution to

$$\begin{cases} u''(x)+\omega^2 u(x)=f(x), & 0<x<l; \\ u'(x)=u(l)=0. \end{cases} \tag{23.14}$$

Problem 23.7. Construct Green's functions and write the solutions for the following boundary value problems:

a. $u''(x)=f(x),\ 0<x<1;\ u'(0)=u(0),\ u'(1)=-u(1).$

b. $u''(x)+u(x)=f(x),\ 0<x<1;\ u'(0)=u(0),\ u'(1)=3u(1).$

c. $x^2 u''(x)+2xu'(x)=f(x),\ 1<x<2;\ u'(1)=0,\ u(2)+5u'(2)=0.$

d. $(3+x^2)u''(x)+2xu'(x)=f(x),\ 0<x<1;\ u'(0)=u(0),\ u(1)=0.$

24 Solvability condition for the boundary value problems

Let us point out that the formula (23.13) for the solution of the Sturm – Liouville problem (23.11) and Green's function (23.12) do not make sense for $\omega l = k\pi, k \in \mathbb{N}$, since then $\sin \omega l = 0$.

Question 24.1. Could we foresee this without solving the problem (23.11)?

Answer. Yes. When $\omega l = k\pi$, problem (23.11) has a nonzero solution $u_0(x)$ for $f(x) \equiv 0$, given by $u_0(x) = \sin \frac{k\pi x}{l}$,

$$\begin{cases} \left(\dfrac{d^2}{dx^2} + \omega^2\right) u_0(x) = 0, & 0 < x < l; \\ u_0(0) = u_0(l) = 0. \end{cases} \tag{24.1}$$

Therefore, the operator $A \equiv \frac{d^2}{dx^2} + \omega^2 : C_0^2[0,l] \to C[0,l]$ is not invertible! It follows that the (left) inverse G does not exist, hence its integral kernel $G(x,y)$ is not defined.

We point out, though, that the absence of the inverse operator to A does not mean that the problem (23.11) does not have solutions for a single $f(x)$!

Question 24.2. Under which conditions on $f(x)$ does the problem (23.11) have solution $u(x)$, and how could this solution be found?

Normal solvability for linear algebraic systems

To answer this question, we need to take a detour into the linear algebra. The thing is, the similar question arises when solving the system

$$Au = f, \tag{24.2}$$

where A is a real $n \times n$ matrix and $f \in \mathbb{R}^n$. The system (24.2) has a (unique) solution $u = A^{-1}f$ if $\det A \neq 0$. If instead $\det A = 0$, then the system (24.2) may not have solutions.

The necessary and sufficient condition on f so that the system (24.2) has a solution is the following orthogonality condition (see [CH53]):

$$f \perp \operatorname{Ker} A^*. \tag{24.3}$$

Here $\operatorname{Ker} A^*$ is the subspace in $\mathbb{R}^n$ that consists of solutions to the adjoint homogeneous system:

$$h \in \operatorname{Ker} A^* \iff A^* h = 0.$$

Thus, (24.3) means that

$$\langle f, h \rangle = 0, \qquad \forall h \in \operatorname{Ker} A^*. \tag{24.4}$$

Let us prove the necessity of conditions (24.3), (24.4). If for a given vector $f \in \mathbb{R}^n$ there is a solution u to the system (24.2), then for each vector $h \in \mathrm{Ker}A^*$

$$\langle f, h \rangle = \langle Au, h \rangle = \langle u, A^*h \rangle = \langle u, 0 \rangle = 0. \tag{24.5}$$

Problem 24.3. Prove the sufficiency of the conditions (24.3), (24.4) for the well-posedness of the system (24.2).

Due to the form of the condition (24.3) we say that the system (24.2) is *normally solvable*.

Let us emphasize the particular case when A is self-adjoint, that is, $A^* = A$. In this case, the normal solvability condition (24.3) takes the form

$$f \perp \mathrm{Ker}A. \tag{24.6}$$

Application to the Sturm – Liouville problem

The formula (11.4) shows that the operator $A = \frac{d^2}{dx^2} + \omega^2$ corresponding to the Sturm – Liouville problem is *formally* self-adjoint. Respectively, we suggest that the condition (24.6) is necessary and sufficient for the solvability of the Sturm – Liouville problem (23.11). We will prove this below (see Problems 24.4 and 24.5).

Let us note that $\mathrm{Ker}A$ is the space of solutions to the problem (24.1):

$$\mathrm{Ker}A = \left\{ C \sin \frac{k\pi x}{l} : C \in \mathbb{R} \right\}.$$

Thus, the orthogonality condition (24.6) takes the following form:

$$\left\langle f(x), \sin \frac{k\pi x}{l} \right\rangle = \int_0^l f(x) \sin \frac{k\pi x}{l} \, dx = 0. \tag{24.7}$$

Problem 24.4. Prove the necessity of the condition (24.7) for the solvability of the problem (23.11) when $\omega = \frac{k\pi}{l}$.

Solution. Analogously to (24.5) and in view of (11.4), we have:

$$\left\langle f(x), \sin \frac{k\pi x}{l} \right\rangle = \left\langle \left(\frac{d^2}{dx^2} + \omega^2 \right) u(x), \sin \frac{k\pi x}{l} \right\rangle$$

$$= \left\langle u(x), \left(\frac{d^2}{dx^2} + \omega^2 \right) \sin \frac{k\pi x}{l} \right\rangle = \langle u(x), 0 \rangle = 0.$$

Problem 24.5. Prove the sufficiency of the condition (24.7).

Solution. Let us take $\omega \to \frac{k\pi}{l}$, but $\omega \neq \frac{k\pi}{l}$. Then the problem (23.11) has the solution (23.13). It turns out that, first of all, the function (23.13) under the condition (24.7) has a limit as $\omega \to \frac{k\pi}{l}$, and, secondly, this limit is the solution to the problem (23.11).

Let us prove the first statement. By the formula (23.13),

$$u(x) = \frac{1}{\omega \sin \omega l}\left(\int_0^x \sin \omega y \sin \omega (x-l) f(y)\, dy + \int_x^l \sin \omega (y-l) \sin \omega x\, f(y)\, dy \right).$$
(24.8)

When $\omega = \frac{k\pi}{l}$, the integrands in both integrals have the same form. We use the identities

$$\begin{cases} \sin \omega y \sin \omega (x-l) = \sin \omega y \sin(\omega x - k\pi) = (-1)^k \sin \omega y \sin \omega x, \\ \sin \omega (y-l) \sin \omega x = \sin(\omega y - k\pi) \sin \omega x = (-1)^k \sin \omega y \sin \omega x. \end{cases}$$

Then the expression in the brackets in (24.8) takes the form

$$(-1)^k \sin \omega x \int_0^l \sin \omega y f(y)\, dy.$$
(24.9)

But when $\omega = \frac{k\pi}{l}$ the integral (24.9) is equal to zero due to the orthogonality condition (24.7)! Therefore, when $\omega \to \frac{k\pi}{l}$, both the numerator and the denominator of expression (24.8) tend to 0, and we obtain the indeterminate form $\frac{0}{0}$.

Problem 24.6. Find the limit of (24.8) as $\omega \to \frac{k\pi}{l}$. (Apply the l'Hospital rule.)

Answer.

$$u(x) = \frac{1}{\omega l \sin k\pi}\left\{ \int_0^x [y\cos \omega y \sin \omega (x-l) + \sin \omega y (x-l)\cos \omega (x-l)] f(y)\, dy \right.$$
$$\left. + \int_x^l [(y-l)\cos \omega (y-l) \sin \omega x + \sin \omega (y-l) x \cos \omega x] f(y)\, dy \right\}. \quad (24.10)$$

Problem 24.7. Prove that the function (24.10) is a solution to the problem (23.11) (valid when $\omega = \frac{k\pi}{l}$ and the condition (24.7) is satisfied!).

Question 24.8. The solution to the problem (23.11) for $\omega = \frac{k\pi}{l}$ is not uniquely defined, since one could add to it $C\sin \frac{k\pi x}{l}$ with any value of C. What is a special feature of the solution (24.10) among all other solutions?

Answer. The formula (24.10) gives a solution to the problem (23.11) which satisfies the condition $\left\langle u(x), \sin \frac{k\pi x}{l} \right\rangle = 0$, that is, $u \perp \mathrm{Ker}\, A$.

Problem 24.9. Find the solvability condition and the solution for the problem (23.14) with

$$\omega = \frac{(k + \frac{1}{2})\pi}{l}, \qquad k = 0, 1, 2, \ldots.$$

25 The Sobolev functional spaces

Let Ω be some region in $\mathbb{R}^n$, and $s = 0, 1, 2, \ldots$.

Definition 25.1.

a. *The Sobolev space $H^s(\Omega)$ consists of all functions $u(x) \in L^2(\Omega)$ which satisfy*

$$\partial_x^\alpha u(x) \in L^2(\Omega), \quad \text{for} \quad |\alpha| \leq s,$$

where the derivatives are understood in the sense of distributions.

b. *The Sobolev norm $\|u\|_s$ in the space $H^s(\Omega)$ is defined by*

$$\|u\|_s^2 \equiv \sum_{|\alpha| \leq s} \|\partial_x^\alpha u(x)\|_{L^2(\Omega)}^2 = \sum_{|\alpha| \leq s} \int_\Omega |\partial_x^\alpha u(x)|^2 \, dx.$$

Remark 25.2. $H^0(\Omega) \equiv L^2(\Omega)$, and, obviously, $C_0^\infty(\Omega) \subset H^s(\Omega)$.

Definition 25.3. $H_0^s(\Omega)$ *is the closure of $C_0^\infty(\Omega)$ in the space $H^s(\Omega)$.*

Let us list the most important properties of the Sobolev spaces.

Property 25.4. $H^s(\Omega)$ is the complete Hilbert space. Later we will always assume that Ω is a bounded region in $\mathbb{R}^n$ (with the compact closure $\bar{\Omega}$) and the smooth boundary $\partial\Omega$.

The two following properties are known as *Sobolev embedding theorems*:

Property 25.5. $H^s(\Omega) \subset C(\bar{\Omega})$ for $s > \frac{n}{2}$.

Property 25.6. For $s_1 > s_2$, the inclusion $H^{s_1}(\Omega) \subset H^{s_2}(\Omega)$ is compact.

Proofs of Properties 25.4, 25.5, and 25.6 are in [Pet91, SD64].

Let us consider examples of the Sobolev spaces. Let $n = 1$ and $\Omega = (0,l)$, where $l > 0$. Then:

a. $H^0(0,l) = L^2(0,l)$, and, decomposing $u(x) \in L^2(0,l)$ into the Fourier series $u(x) = \sum_{k=1}^\infty u_k \sin \frac{k\pi x}{l}$ and applying the Bessel identity, we get:

$$\|u\|_0^2 = \frac{l}{2} \sum_{k=1}^\infty |u_k|^2. \tag{25.1}$$

b. The space $H^1(0,l)$ consists of functions $u(x) \in L^2(0,l)$ which satisfy

$$\|u\|_1^2 \equiv \int_0^l |u(x)|^2 \, dx + \int_0^l |u'(x)|^2 \, dx < \infty. \tag{25.2}$$

c. The space $H_0^1(0,l)$ consists of functions $u(x) \in C[0,l]$ such that norm (25.2) is finite, and, moreover, $u(0) = u(l) = 0$. (Prove this!)

Problem 25.7. Prove that, analogously to (25.1), for $u \in H_0^1(0,l)$

$$\|u\|_1^2 = \frac{l}{2}\sum_{k=1}^{\infty}|u_k|^2 + \frac{\pi^2}{2l}\sum_{k=1}^{\infty}k^2|u_k|^2. \qquad (25.3)$$

Hint. First prove (25.3) in the case $u(x) \in C_0^{\infty}(0,l)$.

Corollary 25.8. For $u \in H_0^1(0,l)$ the norm $\|u\|_1^2$ is equivalent to the norm

$$|||u|||_1^2 \equiv \sum_{k=1}^{\infty}k^2|u_k|^2.$$

Problem 25.9. Prove that (25.3) is not valid for $u(x) \in H^1(0,l) \setminus H_0^1(0,l)$.

Problem 25.10. Prove the *Poincaré inequality*: For $u(x) \in H_0^1(0,l)$,

$$\int_0^l |u(x)|^2\,dx \leq C\int_0^l |u'(x)|^2\,dx, \qquad (25.4)$$

where $C > 0$ does not depend on u.

Hint. Express the integrals in (25.4) via the Fourier coefficients u_k with respect to the basis

$$\left\{\sin\frac{k\pi x}{l} : k \in \mathbb{N}\right\}.$$

Problem 25.11. Prove that there is no constant $C > 0$ such that the inequality (25.4) holds for all $u \in H^1(0,l)$.

Problem 25.12. Prove that $H_0^1(0,l) \subset C[0,l]$, using the finiteness of the norm (25.3) and decomposition of $u(x)$ into the Fourier series.

Problem 25.13. Prove the compactness of the embedding

$$H_0^1(0,l) \subset H^0(0,l)$$

using (25.1) and (25.3).

Problem 25.14. Let Ω be a ball $|x| < 1$ in $\mathbb{R}^n$, $n \geq 1$. For which $\alpha \in \mathbb{R}$ do the the following inclusions take place:

a. $|x|^{\alpha} \in H^1(\Omega)$?

b. $(\sin|x|)^{\alpha} \in H^1(\Omega)$?

c. $(\ln|x|)^{\alpha} \in H^1(\Omega)$?

26 Well-posedness of the wave equation in the Sobolev spaces

Consider the problem (14.1) and the formula (14.6) for its solution. Assume that

$$\varphi(x) \in H_0^1(0,l) \quad \text{and} \quad \psi(x) \in H^0(0,l) \equiv L^2(0,l). \tag{26.1}$$

Let us verify that the formula (14.6) with the initial data φ, ψ from (26.1) gives the solution to the problem (14.1).

Problem 26.1. Prove that the function (14.6) satisfies equation (14.1) in the sense of distributions from $\mathscr{D}'((0,l) \times \mathbb{R})$.

Problem 26.2. Prove that for $\forall t \in \mathbb{R}$

$$u(x,t) \in H_0^1(0,l) \quad \text{and} \quad \dot{u}(x,t) \in H^0(0,l).$$

Problem 26.3. Prove that the mapping $t \mapsto u(x,t)$ is continuous from $\mathbb{R}$ into $H_0^1(0,l)$, while $t \mapsto \dot{u}(x,t)$ is continuous from $\mathbb{R}$ to $H^0(0,l)$, and, moreover, that

$$u(\cdot,t) \xrightarrow{\ H_0^1\ } \varphi(\cdot) \quad \text{and} \quad \dot{u}(\cdot,t) \xrightarrow{\ H^0\ } \psi(\cdot) \quad \text{as} \quad t \to 0.$$

Problem 26.4. Prove the uniqueness of a solution to the problem (14.1) in the class of functions $u(x,t)$ which possess properties formulated in Problems 26.1, 26.2, and 26.3.

Problem 26.5. For $t \in \mathbb{R}$, denote by W_t the mapping

$$W_t : (\varphi, \psi) \longmapsto (u(\cdot,t), \dot{u}(\cdot,t))$$

represented by the formula (14.6). According to the statement of Problem 26.2, the mapping W_t maps the space $E \equiv H_0^1(0,l) \times H^0(0,l)$ into itself.

Problem 26.6. Prove that for all $t \in \mathbb{R}$ the mapping $W_t : E \to E$ is continuous.

Problem 26.7. Prove that the mapping W_t defined above form a group:

$$W_s W_t = W_{s+t}, \qquad \forall s, t \in \mathbb{R}.$$

Problem 26.8. Prove the energy conservation for the solution (14.6):

$$H \equiv \int_{-\infty}^{\infty} \left(|\dot{u}(x,t)|^2 + |u'(x,t)|^2 \right) dx = \text{const}, \qquad t \in \mathbb{R}.$$

Remark 26.9. It is for justification of the solution (14.6) under assumptions (26.1) that S.L. Sobolev introduced the functional spaces H^s. The Sobolev theory gives a mathematically rigorous approach to the analysis of all *finite energy solutions*.

27 Solutions to the wave equation in the sense of distributions

As we have seen in Section 2, a solution to the homogeneous d'Alembert equation

$$\frac{\partial^2 u(x,t)}{\partial t^2} = a^2 \frac{\partial^2 u(x,t)}{\partial x^2} \tag{27.1}$$

can be written in the form (2.3):

$$u(x,t) = f(x - at) + g(x + at). \tag{27.2}$$

If the functions $f(x)$ and $g(x)$ are C^2 (have two continuous derivatives), then $u(x,t)$ from (27.2) also possesses the same property. If, instead, $f(x)$ or $g(x)$ are discontinuous, then so is $u(x,t)$. Let us show that in this case the function $u(x,t)$ in (27.2) is still a solution to equation (27.1) if one considers the derivatives in both sides of (27.1) in the sense of distributions (see Remark 2.1). This means that

$$\left\langle \frac{\partial^2 u}{\partial t^2}, \varphi(x,t) \right\rangle = a^2 \left\langle \frac{\partial^2 u}{\partial x^2}, \varphi(x,t) \right\rangle, \qquad \forall \varphi \in C_0^\infty(\mathbb{R}^2). \tag{27.3}$$

To prove the identity (27.3), let us remind that, according to the definition of the derivatives of distributions,

$$\left\langle \frac{\partial^2 u}{\partial t^2}, \varphi \right\rangle = -\left\langle \frac{\partial u}{\partial t}, \frac{\partial \varphi}{\partial t} \right\rangle = \left\langle u, \frac{\partial^2 \varphi}{\partial t^2} \right\rangle, \qquad \left\langle \frac{\partial^2 u}{\partial x^2}, \varphi \right\rangle = -\left\langle \frac{\partial u}{\partial x}, \frac{\partial \varphi}{\partial x} \right\rangle = \left\langle u, \frac{\partial^2 \varphi}{\partial x^2} \right\rangle.$$

Therefore, the identity (27.3) is equivalent to $\left\langle u, \frac{\partial^2 \varphi}{\partial t^2} \right\rangle = a^2 \left\langle u, \frac{\partial^2 \varphi}{\partial x^2} \right\rangle$, or

$$\left\langle u, \frac{\partial^2 \varphi}{\partial t^2} - a^2 \frac{\partial^2 \varphi}{\partial x^2} \right\rangle = 0. \tag{27.4}$$

The identity (27.4) in the coordinates $\xi = x - at$, $\eta = x + at$ takes the following form (Cf. (2.7)):

$$\int_{-\infty}^{+\infty} \int_{-\infty}^{+\infty} u(\xi, \eta) \frac{\partial^2 \varphi}{\partial \xi \partial \eta} d\xi \, d\eta = 0. \tag{27.5}$$

According to (27.2), $u(\xi, \eta) = f(\xi) + g(\eta)$. Substituting this expression into (27.5), we get:

$$\int_{-\infty}^{+\infty} f(\xi) \left(\int_{-\infty}^{+\infty} \frac{\partial^2 \varphi}{\partial \xi \partial \eta} d\eta \right) d\xi + \int_{-\infty}^{+\infty} g(\eta) \left(\int_{-\infty}^{+\infty} \frac{\partial^2 \varphi}{\partial \xi \partial \eta} d\xi \right) d\eta = 0.$$

But this equality is immediate due to the compact support of $\varphi(\xi, \eta)$:

$$\int_{-\infty}^{+\infty} \frac{\partial^2 \varphi}{\partial \xi \partial \eta} d\eta = \frac{\partial \varphi}{\partial \xi}(\xi, \eta) \Big|_{\eta = -\infty}^{\eta = +\infty} = 0, \qquad \int_{-\infty}^{+\infty} \frac{\partial^2 \varphi}{\partial \xi \partial \eta} d\xi = \frac{\partial \varphi}{\partial \eta}(\xi, \eta) \Big|_{\xi = -\infty}^{\xi = +\infty} = 0.$$

Thus, the equality (27.3) is proved.

Chapter 4
Fundamental solutions and Green's functions in higher dimensions

28 Fundamental solutions of the Laplace operator in $\mathbb{R}^n$

Distributions of several variables $x_1, \ldots, x_n$ and operations with them are defined similarly to the case $n = 1$ (see Sections 19, 20).

a. For example, the Dirac δ-function in $\mathbb{R}^n$ is defined by

$$\langle \delta_{(n)}, \varphi \rangle \equiv \varphi(0), \qquad \forall \varphi \in C_0^\infty(\mathbb{R}^n); \tag{28.1}$$

b. For each distribution $u(x) \in \mathscr{D}'(\mathbb{R}^n)$,

$$\left\langle \frac{\partial u}{\partial x_2}, \varphi \right\rangle \equiv -\left\langle u, \frac{\partial \varphi}{\partial x_2} \right\rangle, \qquad \forall \varphi \in C_0^\infty(\mathbb{R}^n). \tag{28.2}$$

Denote the Laplace operator in $\mathbb{R}^n$ by

$$\triangle_n = \frac{\partial^2}{\partial x_1^2} + \ldots + \frac{\partial^2}{\partial x_n^2}. \tag{28.3}$$

Usually one simply writes $\triangle$ for $\triangle_n$.

In this section we will find the fundamental solution of the operator $\triangle_3$, that is, the function $\mathscr{E}(x) \in \mathscr{D}'(\mathbb{R}^3)$ such that

$$\triangle_3 \mathscr{E}(x) = \delta_{(3)}(x), \qquad x \in \mathbb{R}^3. \tag{28.4}$$

First, we approximate $\delta_{(3)}$-function by the step-functions $\delta_\varepsilon(x)$, analogous to the Steklov step-functions (20.8):

$$\delta_\varepsilon(x) = \begin{cases} \dfrac{1}{\Omega_\varepsilon}, & |x| < \varepsilon, \\[2mm] 0, & |x| > \varepsilon, \end{cases} \tag{28.5}$$

Alexander Komech and Andrew Komech, *Principles of Partial Differential Equations*, Problem Books in Mathematics, DOI·10.2007/978-1-4419-1096-7_4, © Springer Science + Business Media, LLC 2009

where $\Omega_\varepsilon = \frac{4}{3}\pi\varepsilon^3$ is the volume of a ball of radius $\varepsilon > 0$.

Problem 28.1. Prove that, analogously to (20.9),

$$\delta_\varepsilon(x) \xrightarrow{\mathscr{D}'(\mathbb{R}^3)} \delta_{(3)}(x) \quad \text{as} \quad \varepsilon \to 0+. \tag{28.6}$$

We find the solution to equation (28.4) as the limit as $\varepsilon \to 0+$ of solutions $\mathscr{E}_\varepsilon$ to the equation

$$\triangle_3 \mathscr{E}_\varepsilon(x) = \delta_\varepsilon(x), \quad x \in \mathbb{R}^3. \tag{28.7}$$

It is natural to look for the solution to this equation in the form $\mathscr{E}_\varepsilon(x) \equiv E_\varepsilon(r)$, where $r = |x|$. To accomplish this, we introduce the spherical coordinates in (28.7).

Problem 28.2. Prove that for any smooth function which depends on $|x|$ only,

$$\triangle_3\varphi(r) = \frac{\partial^2\varphi}{\partial r^2} + \frac{2}{r}\frac{\partial\varphi}{\partial r} = \frac{1}{r}\frac{\partial^2}{\partial r^2}(r\varphi). \tag{28.8}$$

Corollary 28.3. As it follows from equation (28.7),

$$\frac{1}{r}(rE_\varepsilon)''_{rr} = \delta_\varepsilon(r) \quad \text{for} \quad r > 0, \tag{28.9}$$

where

$$\delta_\varepsilon(r) \equiv \begin{cases} \dfrac{1}{\Omega_\varepsilon}, & 0 < r < \varepsilon; \\[2mm] 0, & r > \varepsilon. \end{cases}$$

Equation (28.9) is readily solved by integrating twice:

$$E_\varepsilon(r) = \begin{cases} \dfrac{r^2}{6}\dfrac{1}{\Omega_\varepsilon} + \dfrac{C_1}{r} + C_2, & 0 < r < \varepsilon; \\[3mm] \dfrac{C_3}{r} + C_4, & r > \varepsilon. \end{cases} \tag{28.10}$$

It remains to find constants $C_1, \ldots, C_4$.

We take into account that, according to (28.9), E_ε and E'_ε are continuous at $\varepsilon = r$ (this follows from the formula (21.1)):

$$\begin{cases} \dfrac{\varepsilon^2}{6}\dfrac{1}{\Omega_\varepsilon} + \dfrac{C_1}{\varepsilon} + C_2 = \dfrac{C_3}{\varepsilon} + C_4, \\[3mm] \dfrac{\varepsilon}{3}\dfrac{1}{\Omega_\varepsilon} - \dfrac{C_1}{\varepsilon^2} = -\dfrac{C_3}{\varepsilon^2}. \end{cases} \tag{28.11}$$

We choose C_1 so that $E_\varepsilon(0+)$ is finite, thus setting $C_1 = 0$. Then the second equation in (28.11) gives

$$C_3 = -\frac{\varepsilon^3}{3\Omega_\varepsilon} = -\frac{1}{4\pi}. \tag{28.12}$$

Since one can add an arbitrary constant to a solution of equation (28.7), we can take $C_4 = 0$. We then have

$$\lim_{r\to\infty} E_{\varepsilon}(r) = 0.$$

Using the value of C_3 from (28.11), we derive from the first equation in (28.11) that

$$C_2 = \frac{C_3}{\varepsilon} - \frac{\varepsilon^2}{6\Omega_{\varepsilon}} = -\frac{1}{4\pi\varepsilon} - \frac{1}{8\pi\varepsilon} = -\frac{3}{8\pi\varepsilon}. \tag{28.13}$$

Thus, $E_{\varepsilon}(r)$ is given by (28.10) with $C_1 = C_4 = 0$ and with the values of C_2 and C_3 from (28.13), (28.12):

$$E_{\varepsilon}(r) = \begin{cases} \dfrac{r^2}{8\varepsilon^3} - \dfrac{3}{8\varepsilon}, & 0 < r < \varepsilon; \\[3mm] -\dfrac{1}{4\pi r}, & r > \varepsilon. \end{cases} \tag{28.14}$$

See Fig. 28.1.

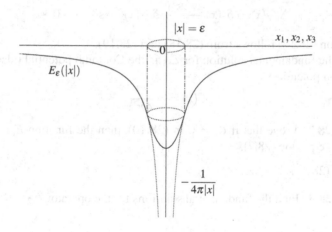

Fig. 28.1

Remark 28.4. The condition that $E_{\varepsilon}(0)$ is finite resulted in $\mathscr{E}_{\varepsilon}(x) \equiv E_{\varepsilon}(|x|)$ being a smooth function in the ball $|x| < \varepsilon$:

$$\mathscr{E}_{\varepsilon}(x) = \begin{cases} \dfrac{x_1^2 + x_2^2 + x_3^2}{8\varepsilon^3} - \dfrac{3}{8\varepsilon}, & |x| < \varepsilon; \\[3mm] -\dfrac{1}{4\pi|x|}, & |x| > \varepsilon. \end{cases} \tag{28.15}$$

Thus, it satisfies equation (28.7) not only for $x \in \mathbb{R}^3 \setminus 0$ (as it follows from (28.9)), but also in an open neighborhood of the point $x = 0$, and hence for all $x \in \mathbb{R}^3$.

Problem 28.5. Prove that, in the sense of convergence of distributions in $\mathscr{D}'(\mathbb{R}^3)$,

$$\mathscr{E}_\varepsilon(x) \xrightarrow{\mathscr{D}'(\mathbb{R}^3)} -\frac{1}{4\pi|x|} \quad \text{as} \quad \varepsilon \to 0+. \tag{28.16}$$

Corollary 28.6. As it follows from (28.16),

$$\Delta\left(-\frac{1}{4\pi|x|}\right) = \delta_{(3)}(x), \quad x \in \mathbb{R}^3. \tag{28.17}$$

Indeed, from (28.16), due to the continuity of the operator Δ in $\mathscr{D}'(\mathbb{R}^3)$ (see Lemma 20.1),

$$\Delta\mathscr{E}_\varepsilon(x) \xrightarrow{\mathscr{D}'(\mathbb{R}^3)} \Delta\left(-\frac{1}{4\pi|x|}\right) \quad \text{as} \quad \varepsilon \to 0+. \tag{28.18}$$

But, on the other hand, from (28.6) and (28.7) we also see that

$$\Delta\mathscr{E}_\varepsilon(x) = \delta_\varepsilon(x) \xrightarrow{\mathscr{D}'(\mathbb{R}^3)} \delta_{(3)}(x) \quad \text{as} \quad \varepsilon \to 0+. \tag{28.19}$$

The relation (28.17) follows from (28.18) and (28.19).

Thus, the fundamental solution for Δ_3 is the Coulomb potential (also known as the Newton potential),

$$\mathscr{E}(x) = -\frac{1}{4\pi|x|}. \tag{28.20}$$

Problem 28.7. Prove that if $C_1 \neq 0$ in (28.10), then the function $E_\varepsilon(|x|)$ is not a solution to equation (28.7).

Hint. Use (28.17).

Problem 28.8. Find the fundamental solutions for the operator Δ_2.

Answer.

$$\mathscr{E}(x) = \frac{1}{2\pi} \ln|x|, \quad x \in \mathbb{R}^2. \tag{28.21}$$

Remark 28.9. Both fundamental solutions (28.20) and (28.21) tend to negative infinity as $|x| \to 0$. On the other hand, (28.21) grows as $|x| \to \infty$, while (28.20) remains finite.

Problem 28.10. Find the fundamental solutions for the operators Δ_n, $n > 3$.

Problem 28.11. Find the fundamental solutions for the operators $\Delta_3 \pm k^2$, $k > 0$.

29 Potentials and their properties

Volume potentials

Once one knows the fundamental solutions to the Laplace equation (28.20) and (28.21), one can also find solutions to the nonhomogeneous Laplace equation in $\mathbb{R}^n$ for $n = 2$ and $n = 3$. For example, the solution to the equation

$$\triangle_2 u(x) = f(x), \qquad x \in \mathbb{R}^2 \tag{29.1}$$

is the function

$$u(x) = \frac{1}{2\pi} \int_{\mathbb{R}^2} \ln |x - y| f(y) \, dy, \tag{29.2}$$

which is well-defined for $f(x) \in C(\mathbb{R}^2)$, $f(x) = 0$ for $|x| > $ const. In the same fashion, a solution to the equation

$$\triangle_3 u(x) = f(x), \qquad x \in \mathbb{R}^3 \tag{29.3}$$

is given by

$$u(x) = -\frac{1}{4\pi} \int_{\mathbb{R}^3} \frac{1}{|x - y|} f(y) \, dy. \tag{29.4}$$

Remark 29.1. Integrals of the form (29.4) are called the *Coulomb* (or *Newton*) *volume* potentials. As the matter of fact, in electrostatics, the integral (29.4) up to a scalar factor (which depends on the choice of units) and up to the sign represents the potential of the electric field of the charge with the volume density $f(x)$. Equation (29.3) for the electric potential is called the Poisson equation. It takes the form (29.1) in a particular case when the charge distribution $f(x)$ does not depend on the coordinate x_3. For example, the potential of a uniformly charged straight infinite wire satisfies (29.1).

It follows that the fundamental solution $-\frac{1}{4\pi|x|}$ from (28.20) is the potential of a point charge $+1$ located at the point $x = 0$, since in this case the charge distribution is given by the Dirac δ-function, $f(x) = \delta_{(3)}(x)$.

Surface potentials: single layer potential

The *single layer potential* is a potential of the charge distributed over a surface:

$$u(x) = -\frac{1}{4\pi} \int_S \frac{1}{|x - y|} \sigma(y) \, dS(y). \tag{29.5}$$

Here S is a smooth compact surface in $\mathbb{R}^3$ and $\sigma(y)$ is the surface charge density.

Problem 29.2. Compute the potential of the uniform distribution of charge on a sphere $|x| = R$, with the surface density σ.

Solution.

$$u(x) = -\frac{1}{4\pi} \int\limits_{|y|=R} \frac{\sigma \, dS}{|x-y|} = -\frac{1}{4\pi} \int\limits_0^\pi \int\limits_0^{2\pi} \frac{\sigma R^2 \sin\Theta \, d\varphi \, d\Theta}{\sqrt{R^2 + |x|^2 - 2R|x|\cos\Theta}}. \tag{29.6}$$

Above, Θ, φ are the spherical coordinates of the point y, counted from the vector x,

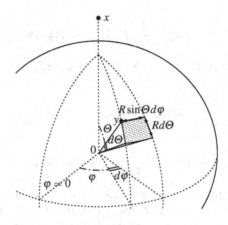

Fig. 29.1

with φ being the longitude and Θ the altitude. See Fig. 29.1. By the cosine theorem,

$$|x-y|^2 = |x|^2 + |y|^2 - 2|x| \cdot |y| \cdot \cos\Theta = |x|^2 + R^2 - 2|x|R\cos\Theta.$$

The integral (29.6) is readily evaluated:

$$u(x) = -\frac{\sigma R^2}{4\pi} 2\pi \int\limits_{\Theta=0}^{\Theta=\pi} \frac{-d(\cos\Theta)}{\sqrt{R^2 + |x|^2 - 2R|x|\cos\Theta}}$$

$$= -\frac{\sigma R^2}{2} \int\limits_1^{-1} \frac{-dt}{\sqrt{R^2 + |x|^2 - 2R|x|t}} = \frac{\sigma R^2}{2} \left[2\frac{\sqrt{R^2 + |x|^2 - 2R|x|t}}{-2R|x|} \right]_{t=1}^{t=-1}$$

$$= -\frac{\sigma R}{2|x|} \left(|R + |x|| - |R - |x|| \right) = \begin{cases} -\sigma R, & |x| \le R; \\ -\dfrac{\sigma R^2}{|x|}, & |x| > R. \end{cases}$$

Answer.

$$u(x) = \begin{cases} -\sigma R, & |x| \leq R; \\ -\dfrac{\sigma R^2}{|x|}, & |x| > R. \end{cases} \qquad (29.7)$$

See Fig. 29.2.

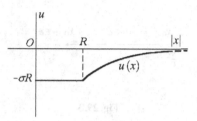

Fig. 29.2

Let us point out that for $|x| > R$ the potential (29.7) coincides with the Coulomb potential u_Q of the point charge of magnitude $Q = 4\pi R^2 \sigma$ equal to the charge of the sphere:

$$u_Q(x) = -\frac{1}{4\pi}\frac{Q}{|x|} = -\frac{1}{4\pi}\frac{4\pi R^2 \sigma}{2|x|} = -\frac{\sigma R}{2|x|}.$$

Remark 29.3. The single layer potential (29.7) is continuous on the sphere $|x| = R$, while its normal derivative is discontinuous, with

$$\left.\frac{\partial u}{\partial n}\right|_{|x|=R+0} - \left.\frac{\partial u}{\partial n}\right|_{|x|=R-0} = \left.\frac{\sigma R^2}{|x|^2}\right|_{|x|=R} = \sigma,$$

where n is the outer normal to the sphere. Besides, one can easily see from (29.7) that

$$\triangle u(x) = 0 \quad \text{for} \quad |x| \neq R.$$

It turns out that these properties of the single layer potential are common for the integrals of the form (29.5).

Properties of the single layer potential:

a. If $\sigma(y)$ is a continuous function, then so is $u(x)$ for all $x \in \mathbb{R}^3$, including $x \in S$;
b. If $\sigma(y)$ has a continuous derivative, then

$$\frac{\partial u}{\partial n}(x+0n) - \frac{\partial u}{\partial n}(x-0n) = \sigma(x),$$

where n is the normal to S at the point $x \in S$;
c. For $x \notin S$, the potential is a harmonic function: $\triangle_3 u(x) = 0$, $x \in \mathbb{R}^3 \setminus S$.

Surface potentials: double layer potential

The *double layer* potential is a potential of the surface distribution of *dipoles*.

First, let us compute the potential of a single dipole, which is a pair of point charges $-\frac{p}{\varepsilon}$ and $+\frac{p}{\varepsilon}$ at an "infinitely small" distance ε from one another, in the direction of a unit vector $\boldsymbol{e}$. The vector $p\boldsymbol{e}$ (see Fig. 29.3) is called *the dipole moment*.

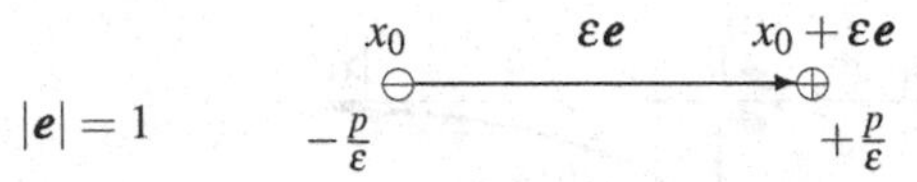

$$|\boldsymbol{e}| = 1 \qquad \begin{array}{ccc} x_0 & \varepsilon\boldsymbol{e} & x_0+\varepsilon\boldsymbol{e} \\ \ominus & \longrightarrow & \oplus \\ -\frac{p}{\varepsilon} & & +\frac{p}{\varepsilon} \end{array}$$

Fig. 29.3

The dipole potential is equal to

$$u(x) = -\lim_{\varepsilon \to 0} \frac{1}{4\pi} \left(\frac{-\frac{p}{\varepsilon}}{|x-x_0|} + \frac{\frac{p}{\varepsilon}}{|x-(x_0+\varepsilon\boldsymbol{e})|} \right)$$

$$= -\frac{p}{4\pi} \frac{d}{d\varepsilon}\bigg|_{\varepsilon=0} \frac{1}{|x-x_0-\varepsilon\boldsymbol{e}|} = -\frac{p}{4\pi} \frac{1}{|x-x_0|^2} \cos(\widehat{x-x_0,\boldsymbol{e}}). \qquad (29.8)$$

The sign of expression (29.8) could be checked by considering the case when the directions of the vectors $x - x_0$ and $\boldsymbol{e}$ coincide.

Let us find the double layer potential on the surface S in $\mathbb{R}^3$ (see Fig. 29.4) with the dipole density $p(y)$, with the dipole moments in the direction of the normal $\boldsymbol{n}_y$ to the surface at every $y \in S$:

$$u(x) = -\frac{1}{4\pi} \int_S \frac{p(y)\cos(\widehat{x-y,\boldsymbol{n}_y})\,dS(y)}{|x-y|^2}. \qquad (29.9)$$

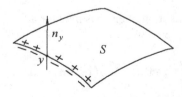

Fig. 29.4

Problem 29.4. Compute the potential of the double layer potential for a sphere with the constant dipole density p.

Solution. We consider the single layer potential for the spheres of radii $R + \varepsilon$ and R with the charge density $\frac{p_\varepsilon}{\varepsilon}$ and $-\frac{p}{\varepsilon}$, respectively (Fig. 29.5).

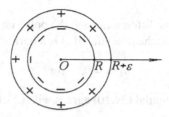

Fig. 29.5

The density p_ε could be determined from the fact that the total charge of the spheres is equal to zero:

$$\frac{p_\varepsilon}{\varepsilon} \cdot 4\pi (R+\varepsilon)^2 - \frac{p}{\varepsilon} 4\pi R^2 = 0,$$

since the sum of charges in each dipole equals zero! Hence,

$$p_\varepsilon = p \frac{r^2}{(R+\varepsilon)^2}.$$

Using the formula (29.7), we obtain the desired double layer potential:

$$u(x) \approx \begin{cases} -\dfrac{p_\varepsilon}{\varepsilon}(R+\varepsilon) + \dfrac{p}{\varepsilon}R, & |x| < R; \\[2mm] -\dfrac{p_\varepsilon(R+\varepsilon)^2}{\varepsilon|x|} + \dfrac{pR^2}{\varepsilon|x|}, & |x| > R+\varepsilon. \end{cases}$$

Taking the limit $\varepsilon \to 0$, we obtain the exact formula

$$u(x) = \begin{cases} -\dfrac{d}{d\varepsilon}\bigg|_{\varepsilon=0}(R+\varepsilon)p_\varepsilon, & |x| < R; \\[2mm] -\dfrac{1}{|x|}\dfrac{d}{d\varepsilon}\bigg|_{\varepsilon=0}p_\varepsilon(R+\varepsilon)^2, & |x| > R. \end{cases}$$

Answer.

$$u(x) = \begin{cases} p, & |x| < R; \\ 0, & |x| > R. \end{cases} \tag{29.10}$$

Properties of the double layer potential:

a. The double layer potential (29.9) is a function which is discontinuous at the points of the surface S:

$$u(x+0 \cdot n_x) - u(x-0 \cdot n_x) = -p(x), \quad x \in S \qquad (29.11)$$

(if the function $u(x)$ is differentiable at the point x);

b. Beyond the surface S, the potential $u(x)$ is a harmonic function:

$$\triangle_3 u(x) = 0, \quad \text{for} \quad x \in \mathbb{R}^3 \setminus S. \qquad (29.12)$$

Example 29.5. The potential (29.10) agrees with (29.11) and (29.12).

Computation of volume potentials

Problem 29.6. Compute the potential of the uniform distribution of charges with the density ρ in a spherical layer $R_1 < |x| < R_2$.

Solution. The potential we are looking for can be converted to a form

$$u(x) = -\frac{1}{4\pi} \int_{R_1 < |y| < R_2} \frac{\rho \, dy}{|x-y|} = \int_{R_1}^{R_2} u_r(x) \, dr. \qquad (29.13)$$

Here $u_r(x)$ is the potential of the same form as in (29.6), obtained according to (29.7):

$$u_r(x) = -\frac{1}{4\pi} \int_{|y|=r} \frac{\rho \, dS(y)}{|x-y|} = \begin{cases} -\rho r, & |x| < r; \\ -\dfrac{\rho r^2}{|x|}, & |x| > r. \end{cases}$$

We consider the following three cases:

a. For $|x| < R_1$, $u(x) = \int_{R_1}^{R_2}(-\rho r)\, dr = -\rho\left(\frac{R_2^2}{2} - \frac{R_1^2}{2}\right)$;

b. For $R_1 < |x| < R_2$,

$$u(x) = \int_{R_1}^{|x|}\left(-\frac{\rho r^2}{|x|}\right) dr + \int_{|x|}^{R_2}(-\rho r)\, dr = -\frac{\rho}{|x|}\left(\frac{|x|^3}{3} - \frac{R_1^3}{3}\right) - \rho\left(\frac{R_2^2}{2} - \frac{|x|^2}{2}\right);$$

c. For $|x| > R_2$,

$$u(x) = \int_{R_1}^{R_2}\left(-\frac{\rho r^2}{|x|}\right) dr = -\frac{\rho}{|x|}\left(\frac{R_2^3}{3} - \frac{R_1^3}{3}\right). \qquad (29.14)$$

The graph of the potential (29.13) is plotted on Fig. 29.6.

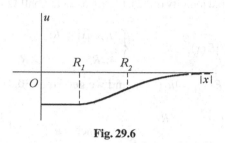

Fig. 29.6

Remark 29.7. For $|x| > R_2$, the potential (29.14) is equal to the Coulomb potential of a point charge, of the value equal to the total charge of the spherical layer:

$$u(x) = -\frac{1}{4\pi} \frac{\frac{4}{3}\pi(R_2^3 - R_1^3)}{|x|}.$$

30 Computing potentials via the Gauss theorem

Since $\triangle u = \operatorname{div}\operatorname{grad} u$, the Poisson equation (29.3) can be written in the form

$$\operatorname{div}\operatorname{grad} u(x) = f(x).$$

Integrating this relation over an arbitrary domain $\Omega \subset \mathbb{R}^3$ and using the Gauss theorem (sometimes called the divergence theorem), we obtain:

$$\int_{\partial\Omega} \operatorname{grad} u(x) \cdot n_x \, dS(x) = \int_{\Omega} \operatorname{div}\operatorname{grad} u(x) \, dx = \int_{\Omega} f(x) \, dx. \qquad (30.1)$$

In electrostatics, $\varphi(x) = -u(x)$ is the potential of the charge distribution $f(x)$, $E(x) = -\operatorname{grad}\varphi(x) = \operatorname{grad} u(x)$ is the intensity vector of the electric field at the point x, and $Q(\Omega) = \int_{\Omega} f(x) \, dx$ is the total charge in the region Ω. Hence, (30.1) could be written in the form

$$\int_{\partial\Omega} E(x) \cdot n_x \, dS(x) = Q(\Omega). \qquad (30.2)$$

This identity is valid for any region $\Omega \subset \mathbb{R}^3$.

Let us compute the potential (29.6) with the aid of the Gauss theorem. The charge density in (29.6) is spherically symmetric, hence the potential $u(x)$ also possesses this property. Therefore, $u(x) = u_1(|x|)$, for some function u_1. Thus, the field $E(x) = -\operatorname{grad} u(x)$ is radial:

$$E(x) = -\frac{x}{|x|} u_1'(|x|). \qquad (30.3)$$

Applying to this field the identity (30.2), taking as Ω the ball $\Omega = \{x \in \mathbb{R}^3 : |x| < r\}$, we get:

$$|E(x)| \cdot 4\pi |x|^2 = \begin{cases} 0, & |x| < R; \\ 4\pi R^2 \sigma, & |x| > R. \end{cases} \tag{30.4}$$

According to (30.3), $|E(x)| = |u'_1(|x|)|$, and we get from (30.4):

$$u'_1(r) \cdot 4\pi r^2 = \begin{cases} 0, & r < R; \\ 4\pi R^2 \sigma, & r > R, \end{cases} \quad \text{hence} \quad u'_1(r) = \begin{cases} 0, & r < R; \\ \frac{R^2 \sigma}{r^2}, & r > R. \end{cases}$$

Integrating, we obtain:

$$u_1(r) = \begin{cases} C_1, & r < R; \\ -\frac{R^2 \sigma}{r} + C_2, & r > R. \end{cases} \tag{30.5}$$

Problem 30.1. Derive the formula (29.7) from (30.5).

Hint. Constants C_1 and C_2 are determined from the continuity condition at $r = R$ and from the equality $\lim_{|x| \to \infty} u(x) = 0$, which easily follows from (29.6).

Problem 30.2. Use the Gauss theorem to compute the potential (29.13).

31 Method of reflections

The Dirichlet problem with zero boundary conditions is solved using the method of odd reflections, while the Neumann problem is solved using the method of even reflections. This is analogous to the method of odd and even extentions from Sections 5 and 6.

The Dirichlet problem in the half-space

Let us illustrate the method of reflections by solving the Dirichlet problem for the Laplace equation in the half-space $\mathbb{R}^3_+ = \{x \in \mathbb{R}^3 : x_3 > 0\}$:

$$\begin{cases} \triangle_3 u(x) = f(x_1, x_2, x_3), & x_1, x_2 \in \mathbb{R}, \quad x_3 > 0; \\ u|_{x_3=0} = 0, & u(x) \xrightarrow[|x| \to \infty]{} 0. \end{cases} \tag{31.1}$$

Here $f(x)$ is a given function in $\mathbb{R}^3_+$, $f(x) \in C(\bar{\mathbb{R}}^3_+)$, $f(x) \equiv 0$ for $|x| > \text{const}$.

Let us find Green's function $G(x,y)$ for the problem (31.1). By definition (compare with (23.2)), G is a solution to the problem

$$\begin{cases} \triangle_x G(x,y) = \delta(x-y), & x_3 > 0; \\ G((x_1,x_2,0),y) = 0; & G(x,y) \to 0 \quad \text{as} \quad |x| \to \infty; \end{cases} \tag{31.2}$$

$G(x,y)$ is smooth for $x \neq y$. Here y is an arbitrary fixed point from $\mathbb{R}^3_+$.

Denote by $\bar{y} = (y_1, y_2, -y_3)$ the point symmetric to y with respect to the boundary $x_3 = 0$ of the half-space $\mathbb{R}^3_+$ (see Fig. 31.1).

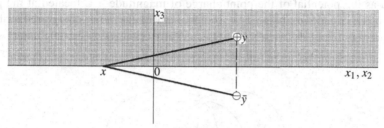

Fig. 31.1

Then the solution to the problem (31.2) is given by the function

$$G(x,y) = -\frac{1}{4\pi}\frac{1}{|x-y|} + \frac{1}{4\pi}\frac{1}{|x-\bar{y}|}. \tag{31.3}$$

According to (28.17), $\triangle_x G(x,y) = \delta(x-y) - \delta(x-\bar{y})$. This yields the first equation in (31.2), since

$$\delta(x-\bar{y}) = 0 \quad \text{for} \quad x_3 > 0.$$

Indeed, $\delta(x-\bar{y})$ is a distribution supported at the point $\bar{y}$ of the lower half-space!

Let us verify the boundary condition in (31.2): If $x_3 = 0$, then the distances $|x-y|$ and $|(x-\bar{y})|$ are equal, as one sees on Fig. 31.1. Therefore from (31.3) one obtains $G(x,y) = 0$. Finally, it is clear that $G(x,y) \xrightarrow[|x| \to \infty]{} 0$.

Solution to the boundary value problem (31.1) is given by the integral

$$u(x) = \int_{\mathbb{R}^3_+} G(x,y)f(y)\,dy = -\frac{1}{4\pi}\int_{y_3>0}\left(\frac{1}{|x-y|} - \frac{1}{|x-\bar{y}|}\right)f(y)\,dy.$$

Indeed, in the sense of distributions,

$$\triangle_x u(x) = \int_{\mathbb{R}^3_+} \triangle_x G(x,y)f(y)\,dy = \int_{\mathbb{R}^3_+} \delta(x-y)f(y)\,dy = f(x).$$

The boundary condition is readily verified:

$$u\big|_{x_3=0} = \int_{\mathbb{R}^3_+} G(x,y)\Big|_{x_3=0} f(y)\,dy = 0, \qquad u(x) \xrightarrow[|x| \to \infty]{} 0.$$

Electrostatic interpretation: Reflected charges

In electrostatics, the solution $u(x)$ to the boundary value problem (31.1), up to a sign and a factor that depends on the metric system, is the potential of the electrostatic field generated by the charge density $f(x)$ in the upper half-plane $\mathbb{R}^3_+$, located above the conducting surface $x_3 = 0$ (this could be, for example, the surface of the Earth or the flat tin roof). Electrostatically, Green's function $G(x,y)$ from (31.2) can be viewed as the potential of the point charge of magnitude $+1$, located at the point y above the conducting plane $x_3 = 0$. The field of the point charge redistributes the charges in the plane $x_3 = 0$: It attracts negative charges while repelling positive charges, forcing them to go to infinity. See Fig. 31.2.

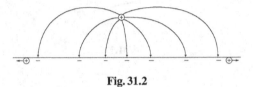

Fig. 31.2

It is known that after this redistribution took place, the field lines (the integral curves) $E(x) = -\operatorname{grad} u(x)$ (see Fig. 31.2) are orthogonal to the conducting surface (the Earth), or else the free charges in the conductor would start moving along the surface. It follows that the surface of a conductor is the level surface of the potential $u(x)$ (equipotential surface in electrostatics). This property of the field lines allows us to find the field $E(x)$. To do this, let us recall the plot of the field curves of the field of two point charges of the same magnitude and opposite sign (see Fig. 31.3).

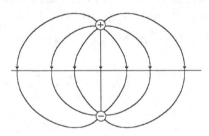

Fig. 31.3

As it follows from the symmetry of the field curves with respect to the plane of symmetry of the charges, the field curves are orthogonal to this plane. Therefore the field above the plane of symmetry coincides with the field we are looking for. This yields the formula (31.3).

The Dirichlet problem in the quarter-space

We consider the quadrant of the space, $\mathbb{R}^3_{++} = \{(x_1, x_2, x_3) \in \mathbb{R}^3 : x_1 \geq 0,\ x_2 \geq 0\}$. Let $f(x) \in C(\mathbb{R}^3_{++})$ be a function such that $f(x) \equiv 0$ for $|x| > C$, with some $C > 0$. Consider the Dirichlet problem in $\mathbb{R}^3_{++}$:

$$\begin{cases} \Delta u(x_1, x_2, x_3) = f(x_1, x_2, x_3), & x_1 > 0,\quad x_2 > 0,\quad x_3 \in \mathbb{R}; \\ u|_{x_1=0} = 0,\quad u|_{x_2=0} = 0; & u(x) \to 0 \quad \text{as} \quad |x| \to \infty. \end{cases} \tag{31.4}$$

Green's function $G(x, y)$ for the problem (31.4), by definition, is a solution of the boundary value problem

$$\begin{cases} \Delta_x G(x, y) = \delta(x - y), & x_1 > 0,\quad x_2 > 0,\quad x_3 \in \mathbb{R}; \\ G|_{x_1=0} = 0,\quad G|_{x_2=0} = 0; & G(x, y) \to 0 \quad \text{as} \quad |x| \to \infty. \end{cases} \tag{31.5}$$

Here $y \in \mathbb{R}^3_{++}$ is a parameter.

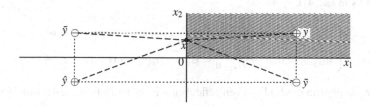

Fig. 31.4

Again, Green's function for this problem can be found by the method of (odd) reflections; see Fig. 31.4. Let $\bar{y} = (y_1, -y_2, y_3)$ be the reflection of the point y in the plane $x_2 = 0$, $\tilde{y} = (-y_1, y_2, y_3)$ be the reflection in the plane $x_1 = 0$, and $\hat{y} = (-y_1, -y_2, y_3)$ be the composition of these two reflections. We put the charges of magnitude $+1$ at the points y and $\hat{y}$, and the charges of magnitude -1 at the points $\bar{y}$ and $\tilde{y}$. Then their electrostatic field is represented by the potential

$$G(x, y) = -\frac{1}{4\pi} \frac{1}{|x - y|} + \frac{1}{4\pi} \frac{1}{|x - \bar{y}|} + \frac{1}{4\pi} \frac{1}{|x - \tilde{y}|} - \frac{1}{4\pi} \frac{1}{|x - \hat{y}|}. \tag{31.6}$$

Let us verify that this function satisfies equation (31.5). First of all, for $x \in \mathbb{R}^3_{++}$,

$$\Delta_x G(x, y) = \delta(x - y) - \delta(x - \bar{y}) - \delta(x - \tilde{y}) + \delta(x - \hat{y}) = \delta(x - y),$$

since $\delta(x - \bar{y})$, $\delta(x - \tilde{y})$, and $\delta(x - \hat{y})$ are equal to zero for $x \in \mathbb{R}^3_{++}$! Therefore, the first equation in (31.5) is satisfied.

Further, let us verify the boundary conditions from (31.5):

a. When $x_1 = 0$, the point x is equidistant from y and $\tilde{y}$, and also from $\bar{y}$ and $\hat{y}$ (see Fig. 31.4). Therefore in the right-hand side of (31.6) the first and the third term cancel out, and so do the second one with the fourth one.
b. Similarly, when $x_2 = 0$, the point x is equidistant from y and $\bar{y}$, and also from $\tilde{y}$ and $\hat{y}$. Therefore, in the right-hand side of (31.6) the first and the second terms cancel out, and so do the third and the fourth ones.
c. At last, it is clear that $G(x,y) \to 0$ as $|x| \to \infty$. Thus, the boundary conditions in (31.5) are also satisfied.

Therefore, $G(x,y)$ from (31.6) is Green's function of the Dirichlet problem (31.4). Hence the solution of the latter could be written in the form

$$u(x) = \int_{\mathbb{R}^3_{++}} G(x,y) f(y)\, dy = -\frac{1}{4\pi} \int_{\mathbb{R}^3_{++}} \left(\frac{1}{|x-y|} - \frac{1}{|x-\bar{y}|} - \frac{1}{|x-\tilde{y}|} + \frac{1}{|x-\hat{y}|} \right) f(y)\, dy.$$

Problem 31.1. Find Green's function and write the formula for the solution to the boundary value problem in a quadrant of the space (under the same conditions on $f(x)$ as in (31.4)):

$$\begin{cases} \Delta u(x) = f(x), & x_1 > 0, \quad x_2 > 0, \quad -\infty < x_3 < \infty; \\ u|_{x_1=0} = 0, \quad \frac{\partial u}{\partial x_2}\big|_{x_2=0} = 0; & u(x) \to 0 \quad \text{as} \quad |x| \to \infty. \end{cases} \tag{31.7}$$

Hint. Apply the method of even reflections in x_1 and of odd reflections in x_2.

Problem 31.2. Find Green's function and write the formula for the solution to the Dirichlet problem in the following domains:

a. A wedge with the dihedral angle $\alpha = \frac{\pi}{n}$, $n = 3, 4, 5, \ldots$.
b. An octant of the three-dimensional space, $x_1 > 0, x_2 > 0, x_3 > 0$.
c. The layer $0 < x_3 < a$; $x_1, x_2 \in \mathbb{R}$ (investigate the convergence of the series).
d. A "half" of the layer: $0 < x_3 < a$, $-\infty < x_1 < \infty$, $x_2 > 0$.
e. A "quarter" of the layer: $0 < x_3 < a$, $x_1 > 0$, $x_2 > 0$.

Remark 31.3. In the previous problem, instead of the Dirichlet condition $u|_{\Gamma} = 0$, one can consider the Neumann condition $\frac{\partial u}{\partial n}\big|_{\Gamma} = 0$ at certain parts of the boundary, as in the problem (31.7). For solutions of such problems one has to use the method of even reflections at these parts of the boundary.

Problem 31.4. Find Green's function of the Dirichlet problem in the following domains:

a. The ball $|x| < R$. *Hint.* Look for Green's function in the form of the sum of the fundamental solution $-\frac{1}{4\pi}\frac{1}{|x-y|}$ and the potential of the "reflected" charge $\frac{q}{4\pi}\frac{1}{|x-y^*|}$ of certain magnitude $q > 0$, located at the point $y^* = \frac{r^2}{|y|^2} y$ which is the *sphere inversion* of the point y.

b. The upper half of the ball $|x| < R$, $x_3 > 0$. *Hint.* Use the method of odd reflections with respect to the plane x_3 to reduce the problem to the ball.

c. A quarter of the ball $|x| < R$, $x_2 > 0$, $x_3 > 0$.

32 Green's functions in 2D via conformal mappings

The Dirichlet problem in half-plane

Let us consider the Dirichlet problem for the Laplace equation in the half-plane $\mathbb{R}^2_+ = \{x \in \mathbb{R}^2 : x_2 > 0\}$:

$$\begin{cases} \Delta u(x_1, x_2) = f(x_1, x_2), & x_1 \in \mathbb{R}, \quad x_2 > 0; \\ u|_{x_2=0} = 0, & u(x) \to 0 \quad \text{as} \quad |x| \to \infty, \end{cases} \tag{32.1}$$

where $f(x) \in C(\overline{\mathbb{R}^2_+})$, $f(x) = 0$ for $|x| > $ const. Green's function $G(x, y)$ of this problem satisfies the equation

$$\begin{cases} \Delta_x G(x, y) = \delta(x - y), & x_1 \in \mathbb{R}, \quad x_2 > 0; \\ G|_{x_2=0} = 0, & G(x, y) \to 0 \quad \text{as} \quad |x| \to \infty, \end{cases}$$

where $y \in \mathbb{R}^2_+$. Similarly to (31.3), this function is found by the method of odd reflections applied to the fundamental solution (28.21) of the Laplace operator in the plane:

$$G(x, y) = \frac{1}{2\pi} \left(\ln|x - y| - \ln|x - \bar{y}| \right), \tag{32.2}$$

where $\bar{y} = (y_1, -y_2)$. See Fig. 32.1.

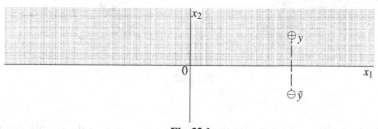

Fig. 32.1

Thus, the solution to the problem (32.1) has the form

$$u(x) = \int_{\mathbb{R}^2_+} G(x, y) f(y) \, dy = \frac{1}{2\pi} \int_{\mathbb{R}^2_+} \ln \frac{|x - y|}{|x - \bar{y}|} f(y) \, dy.$$

The Dirichlet problem for simply connected domains

Unlike in the case of three-dimensional boundary value problems, Green's functions for many simply connected two-dimensional domains could be found with the aid of conformal maps. This is because Green's function $G(x,y)$ is harmonic in x for $x \neq y$, while the conformal mappings send harmonic functions into harmonic ones.

Let us illustrate the relation of Green's functions to the conformal mappings on a particular example of the boundary value problem (32.1). For this, we rewrite the formula (32.2) in the following form:

$$G(x,y) = \frac{1}{2\pi} \ln \frac{|x-y|}{|x-\bar{y}|} = \frac{1}{2\pi} \ln \left| \frac{x-y}{x-\bar{y}} \right|. \qquad (32.3)$$

Remark 32.1. The last equality in (32.3) holds under the condition that $\frac{x-y}{x-\bar{y}}$ is understood in the sense of the division of the complex numbers:

$$\frac{x-y}{x-\bar{y}} = \frac{x_1 + ix_2 - y_1 - iy_2}{x_1 + ix_2 - y_1 + iy_2}.$$

Here $\bar{y} = y_1 - iy_2$ coincides with the complex conjugate of y.

Let us point out that for each fixed $y \in \mathbb{R}^2_+$, the map

$$x \mapsto z = \Phi_y(x) \equiv \frac{x-y}{x-\bar{y}} \qquad (32.4)$$

maps the half-plane $x_2 > 0$ conformally into the unit disc $|z| < 1$ (see Fig. 32.2), and that under the mapping (32.4) the point y is sent to zero: $y \mapsto \Phi_y(y) = 0$.

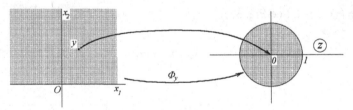

Fig. 32.2

More generally, let us consider the Dirichlet problem in a flat simply connected region $\Omega \subset \mathbb{R}^2$ with a piecewise-smooth boundary $\partial\Omega$, which contains at least two points:

$$\begin{cases} \triangle u(x) = f(x), & x \in \Omega; \\ u|_{x \in \partial\Omega} = 0, \end{cases} \qquad (32.5)$$

where $f(x) \in C(\bar{\Omega})$, $f(x) = 0$ for $|x| > C(y)$. Additionally, if Ω is unbounded, it is required that $u(x) \to 0$ as $|x| \to \infty$, $x \in \Omega$.

Green's function of this problem by definition satisfies the conditions

$$\begin{cases} \triangle_x G(x,y) = \delta(x-y), & x \in \Omega; \\ G\big|_{x\in\partial\Omega} = 0. \end{cases} \tag{32.6}$$

Additionally, if Ω is unbounded, it is required that $G(x,y) \to 0$ as $|x| \to \infty$, $x \in \Omega$. Here $y \in \Omega$ is a parameter. It is known from the theory of functions of a complex variable (the Riemann theorem; see [Ahl66]) that for any simply connected region $\Omega \subset \mathbb{R}^2$ with the boundary $\partial\Omega$ which contains at least two points there exists a conformal mapping of the region Ω onto the unit disc. Moreover, any a priori fixed point y is mapped to zero. Let $\Phi_y(x)$ be such a map (see Fig. 32.3).

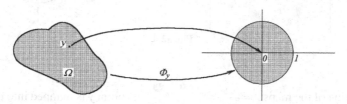

Fig. 32.3

It turns out [CH53] that Green's function (32.6) has the form (32.3):

$$G(x,y) = \frac{1}{2\pi} \ln|\Phi_y(x)|. \tag{32.7}$$

Then we get the solution to the Dirichlet problem (32.5):

$$u(x) = \frac{1}{2\pi} \int_\Omega \ln|\Phi_y(x)| f(y) \, dy. \tag{32.8}$$

Problem 32.2. Check that the function (32.7) is a solution to the problem (32.6).

Hints.

a. $\ln|\Phi_y(x)| = \operatorname{Re} \ln \Phi_y(x)$ is a harmonic function if $\Phi_y(x) \neq 0$, that is, for $x \neq y$.
b. $\ln|\Phi_y(x)|$ at $x = y$ allows the decomposition

$$\ln|\Phi_y(x)| = \ln|x-y| + O(1), \quad x \to y. \tag{32.9}$$

c. Use the theorem about a removable singularity to prove that $O(1)$ in (32.9) is a harmonic function at $x = y$.
d. The boundary condition $G\big|_{\partial\Omega} = 0$ is obviously satisfied, since $\big|\Phi_y(x)\big|_{x\in\partial\Omega}\big| = 1$.

Problem 32.3. Let us find Green's function and the formula for the solution of the Dirichlet problem in the strip Ω (for $f(x) \in C(\bar{\Omega})$, $f(x) = 0$ for $|x| > \text{const}$):

$$\begin{cases} \triangle u(x) = f(x), & 0 < x_2 < a, \quad -\infty < x_1 < \infty; \\ u|_{x_2=0,a} = 0, & u(x) \to 0 \quad \text{when} \quad |x| \to \infty. \end{cases} \quad (32.10)$$

Solution. Let us map conformally the strip Ω into a disc (see Fig. 32.4). By the

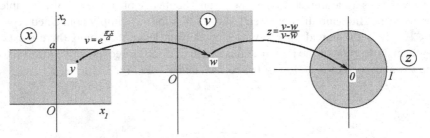

Fig. 32.4

composition of the maps $v = e^{\frac{\pi x}{a}}$, $z = \dfrac{e^{\frac{\pi x}{a}} - e^{\frac{\pi y}{a}}}{e^{\frac{\pi x}{a}} - e^{\frac{\pi \bar{y}}{a}}}$, the point y is mapped into the origin:

$$y \longmapsto w = e^{\frac{\pi y}{a}} \longmapsto 0.$$

According to (32.7), $G(x,y) = \dfrac{1}{2\pi} \ln \left| \dfrac{e^{\frac{\pi x}{a}} - e^{\frac{\pi y}{a}}}{e^{\frac{\pi x}{a}} - e^{\frac{\pi \bar{y}}{a}}} \right|$. Now the solution of the problem (32.10) is given by (32.8):

$$u(x) = \dfrac{1}{2\pi} \int_{-\infty}^{+\infty} \int_0^a \ln \left| \dfrac{e^{\frac{\pi x}{a}} - e^{\frac{\pi y}{a}}}{e^{\frac{\pi x}{a}} - e^{\frac{\pi \bar{y}}{a}}} \right| f(y) \, dy_2 \, dy_1.$$

Let us mention that $e^{\frac{\pi x}{a}} = e^{\frac{\pi}{a}(x_1 + ix_2)} = e^{\frac{\pi x_1}{a}} (\cos \frac{\pi}{a} x_2 + i \sin \frac{\pi}{a} x_2)$.

Problem 32.4. Find Green's function and write the formula for the solution of the Dirichlet problem in the following regions:

a. The angle of magnitude α (see Fig. 32.5).
b. The disc $|x| < 1$ (one gets the classical Poisson formula).
c. A half of the disc: $|x| < 1$, $x_2 > 0$ (use the method of odd reflections with respect to x_2 to reduce to the previous problem).
d. Sector of a disc: $|x| < 1$, $0 < \arg x < \alpha$ (see Fig. 32.6).

Remark 32.5. It turns out that, knowing Green's function of the Dirichlet problem in the region Ω, one can solve the homogeneous equations $\triangle u(x) = 0$ in Ω with

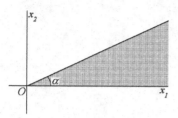

Fig. 32.5

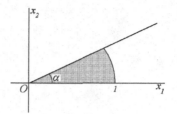

Fig. 32.6

nonhomogeneous boundary conditions $u|_{\partial \Omega} = f(x)$: If $\partial \Omega$ is of class C^2 and $f \in C^2(\partial \Omega)$, then

$$u(x) = \int_{\partial \Omega} \frac{\partial G(x,y)}{\partial n_y} f(y)\, dy,$$

where $\frac{\partial}{\partial n_y}$ is the differentiation in the direction of the external normal to the boundary at the point $y \in \partial \Omega$. This holds for the region Ω of any dimension: $\Omega \subset \mathbb{R}^2$, $\Omega \subset \mathbb{R}^3$, et cetera.

Problem 32.6. Find the solution for the Dirichlet problem in the half-plane:

$$\triangle u(x_1, x_2) = 0 \quad \text{for} \quad x_2 > 0, \qquad u(x_1, 0) = f(x_1), \qquad u(x) \xrightarrow[|x| \to \infty]{} 0,$$

where $f(x_1) \in C(\mathbb{R})$, $f(x_1) = 0$ for $|x_1| > \text{const.}$

Solution. Green's function for the half-plane is given by (32.2):

$$G(x,y) = \frac{1}{2\pi} \ln \sqrt{(x_1 - y_1)^2 + (x_2 - y_2)^2} - \frac{1}{2\pi} \ln \sqrt{(x_1 - y_1)^2 + (x_2 + y_2)^2}$$

$$= \frac{1}{4\pi} \ln\left((x_1 - y_1)^2 + (x_2 - y_2)^2\right) - \frac{1}{4\pi} \ln\left((x_1 - y_1)^2 + (x_2 + y_2)^2\right).$$

Taking into account that the external normal $\boldsymbol{n}_y$ to the half-plane $x_2 > 0$ is represented by the vector $(0, -1)$, we get:

$$\frac{\partial G(x,y)}{\partial n_y}\bigg|_{y_2=0} = -\frac{\partial}{\partial y_2} G(x,y)\bigg|_{y_2=0}$$

$$= \frac{1}{4\pi} \frac{2x_2}{(x_1 - y_1)^2 + x_2^2} + \frac{1}{4\pi} \frac{2x_2}{(x_1 - y_1)^2 + x_2^2} = \frac{1}{\pi} \frac{x_2}{(x_1 - y_1)^2 + x_2^2}.$$

Answer. $u(x_1, x_2) = \dfrac{x_2}{\pi} \displaystyle\int_{-\infty}^{+\infty} \dfrac{f(y_1)\, dy_1}{(x_1 - y_1)^2 + x_2^2}.$

Appendix A
Classification of the second-order equations

Differential equations with constant coefficients

We consider the following equation in $\mathbb{R}^n$:

$$\sum_{i,j=1}^{n} a_{ij} \frac{\partial^2 u}{\partial x_i \partial x_j} + \sum_{i}^{n} a_i \frac{\partial u}{\partial x_i} + a_0 u(x) = 0, \quad x \in \mathbb{R}^n; \quad a_{ij} = a_{ji}, \tag{A.1}$$

where a_{ij}, a_i, and a_0 are constants. Let us bring it to the canonical form, that is, to the form so that $a_{ij} = 0$ for $i \neq j$. To accomplish this, consider the linear change of variables:

$$\begin{cases} y_1 = c_{11}x_1 + \ldots + c_{1n}x_n, \\ \ldots \\ y_n = c_{n1}x_1 + \ldots + c_{nn}x_n, \end{cases} \tag{A.2}$$

or, in the vector form,

$$y = Cx. \tag{A.3}$$

In the coordinates y_k we have

$$\frac{\partial u}{\partial x_i} = \sum_{k=1}^{n} \frac{\partial u}{\partial y_k} \frac{\partial y_k}{\partial x_i} = \sum_{k=1}^{n} C_{ki} \frac{\partial u}{\partial y_k}, \qquad \frac{\partial^2 u}{\partial x_i \partial x_j} = \sum_{k,l=1}^{n} C_{ki} C_{lj} \frac{\partial^2 u}{\partial y_k \partial y_l}.$$

Substituting these identities into (A.1), we get

$$\sum_{i,j,k,l} a_{ij} C_{ki} C_{lj} \frac{\partial^2 u}{\partial y_k \partial y_l} + \ldots = 0, \tag{A.4}$$

where dots denote terms which contain lower order derivatives of the function u. We can write (A.4) in the form

$$\sum_{k,l=1}^{n} b_{kl} \frac{\partial^2 u}{\partial y_k \partial y_l} + \ldots = 0, \tag{A.5}$$

155

where

$$b_{kl} = \sum_{i,j} a_{i,j} C_{kl} C_{lj}.$$

In the matrix form, $b = CaC^*$, where a is the matrix $(a_{ij})_{i,j=1,...,n}$ and C^* is the transpose of C. This formula resembles the transformation law for the matrix of the quadratic form

$$(a\xi,\xi) = \sum_{i,j=1}^{n} a_{ij}\xi_i\xi_j. \tag{A.6}$$

Namely, if one makes the change of variables

$$\xi = d\eta, \quad d = (d_{ij})_{i,j=1,...,n}, \tag{A.7}$$

then, taking $d^* = C$, one gets

$$(a\xi,\xi) = (ad\eta,d\eta) = (d^*ad\eta,\eta) = (CaC^*\eta,\eta) = (b\eta,\eta).$$

Therefore, if the change of variables (A.7) brings the quadratic form to the diagonal form $(a\xi,\xi) = \sum_{k=1}^{n} b_k\eta_k^2$ (we know from linear algebra that such a change of variables exists), then the change of variables (A.3) with the matrix $C = d^*$ brings the differential equation (A.1) to the form (A.5) with the same diagonal matrix b:

$$\sum_{k=1}^{n} b_k \frac{\partial^2 u}{\partial y_k^2} + ... = 0. \tag{A.8}$$

After this is accomplished, one of the following possibilities takes place:

a. $\det a \neq 0$. Then equation (A.1) is called nondegenerate, and all b_k in (A.8) can be made equal to ± 1. Then there are three possibilities:

(*i*) All the coefficients b_k are of the same sign (all are equal to $+1$ or instead all are equal to -1). Then equation (A.8) has the form $\frac{\partial^2 u}{\partial y_1} + ... + \frac{\partial^2 u}{\partial y_n^2} + ... = 0$ and is called *elliptic*. An example is the Laplace equation (8.11).

(*ii*) All the coefficients b_k but one are of the same sign. Then equation (A.8) takes the form

$$\frac{\partial^2 u}{\partial y_k^2} + ... + \frac{\partial^2 u}{\partial y_{n-1}^2} - \frac{\partial^2 u}{\partial y_n^2} + ... = 0 \tag{A.9}$$

and is called *hyperbolic*. An example is the wave equation (7.1).

(*iii*) Some of the coefficients b_k (more than one) are positive, while others (also more than one) are negative. Then equation (A.8) has the form

$$\frac{\partial^2 u}{\partial y_1^2} + \frac{\partial^2 u}{\partial y_2^2} ... - \frac{\partial^2 u}{\partial y_{n-1}^2} - \frac{\partial^2 u}{\partial y_n^2} + ... = 0$$

and is called *ultrahyperbolic*. This is only possible if $n \geq 4$.

b. $\det a = 0$. Then equation (A.1) is called degenerate. An example is the heat equation (8.8).

Problem A.1. Find the canonical form and the corresponding change of variables (A.2) for the equation

$$\frac{\partial^2 u}{\partial x_1^2} + 4\frac{\partial u}{\partial x_1 \partial x_2} - 3\frac{\partial^2 u}{\partial x_3^2} = 0. \tag{A.10}$$

Solution. We write down the quadratic form (A.6) and bring it to the diagonal form:

$$\xi^2 + 4\xi_1\xi_2 - 3\xi_3^2 = (\xi_1 + 2\xi_2)^2 - 4\xi_2^2 - 3\xi_3^2 = \eta_1^2 - \eta_2^2 - 3\eta_3^2. \tag{A.11}$$

Therefore, equation (A.10) is of hyperbolic type, as in (A.9). The change of variables (A.7), or, rather, the inverse to it, has the form

$$\begin{cases} \eta_1 = \xi_1 + 2\xi_2, \\ \eta_2 = 2\xi_2, \\ \eta_3 = \xi_3. \end{cases} \tag{A.12}$$

To bring these relations to the form (A.7), one needs to solve equations (A.12); this yields

$$\begin{cases} \xi_2 = \frac{\eta_2}{2}, \\ \xi_3 = \eta_3, \\ \xi_1 = \eta_1 - 2\xi_2 = \eta_1 - \eta_2. \end{cases}$$

From here we get the matrix d:

$$d = \begin{pmatrix} 1 & -1 & 0 \\ 0 & \frac{1}{2} & 0 \\ 0 & 0 & 1 \end{pmatrix}$$

and, consequently,

$$C = d^* = \begin{pmatrix} 1 & 0 & 0 \\ -1 & \frac{1}{2} & 0 \\ 0 & 0 & 1 \end{pmatrix}$$

Therefore, substitution (A.2) has the form

$$\begin{cases} y_1 = x_1, \\ y_2 = -x_1 + \frac{1}{2}x_2, \\ y_3 = x_3. \end{cases}$$

According to (A.11), the canonical form of equation (A.10) is as follows:

$$\frac{\partial^2 u}{\partial y_1^2} - \frac{\partial^2 u}{\partial y_2^2} - 3\frac{\partial^2 u}{\partial y_3^2} = 0.$$

Problem A.2. Find the canonical form and the change of variables (A.2) for the following equations:

a. $\dfrac{\partial^2 u}{\partial x_1 \partial x_2} + \dfrac{\partial^2 u}{\partial x_2 \partial x_3} + \dfrac{\partial^2 u}{\partial x_3 \partial x_1} = 0;$

b. $\dfrac{\partial^2 u}{\partial x_1^2} - \dfrac{\partial^2 u}{\partial x_1 \partial x_2} + 6\dfrac{\partial^2 u}{\partial x_1 \partial x_3} = 0.$

Equations with variable coefficients

Now we assume that the coefficients in (A.1) are variable:

$$\sum_{i,j=1}^{n} a_{ij}(x) \frac{\partial^2 u(x)}{\partial x_i \partial x_j} + \ldots = 0. \tag{A.13}$$

Then for each fixed $x_0 \in \mathbb{R}^n$ one can consider the equation with the constant coefficients, obtained from the variable coefficients "frozen" at the point x_0:

$$\sum_{i,j=1}^{n} a_{ij}(x_0) \frac{\partial^2 u(x)}{\partial x_i \partial x_j} + \ldots = 0.$$

The type of this equation is called the type of equation (A.13) at the point x_0. The example is the Euler-Tricomi equation

$$\frac{\partial^2 u}{\partial x^2} + x\frac{\partial^2 u}{\partial y^2} = 0,$$

which is elliptic in the half-plane $x > 0$, hyperbolic in the half-plane $x < 0$, and degenerate on the line $x = 0$.

References

[Ahl66] L. V. Ahlfors, *Complex analysis: An introduction of the theory of analytic functions of one complex variable*, Second edition, McGraw-Hill, New York, 1966.

[BJS79] L. Bers, F. John, and M. Schechter, *Partial differential equations*, American Mathematical Society, Providence, R.I., 1979, with supplements by Lars Gårding and A. N. Milgram, With a preface by A. S. Householder, Reprint of the 1964 original, Lectures in Applied Mathematics, 3A.

[CH53] R. Courant and D. Hilbert, *Methods of mathematical physics. Vol. I*, Interscience Publishers, Inc., New York, 1953.

[EKS99] Y. V. Egorov, A. I. Komech, and M. A. Shubin, *Elements of the modern theory of partial differential equations*, Springer-Verlag, Berlin, 1999, translated from the 1988 Russian original by P. C. Sinha, Reprint of the original English edition from the series Encyclopaedia of Mathematical Sciences [*Partial differential equations. II*, Encyclopaedia Math. Sci., 31, Springer, Berlin, 1994].

[Eva98] L. C. Evans, *Partial differential equations*, vol. 19 of *Graduate Studies in Mathematics*, American Mathematical Society, Providence, RI, 1998.

[Hab03] R. Haberman, *Applied partial differential equations*, Prentice Hall Inc., Englewood Cliffs, NJ, 2003, fourth edn., with Fourier series and boundary value problems.

[Pet91] I. G. Petrovsky, *Lectures on partial differential equations*, Dover Publications Inc., New York, 1991, translated from Russian by A. Shenitzer, Reprint of the 1964 English translation.

[Sch66a] L. Schwartz, *Mathematics for the physical sciences*, Hermann, Paris, 1966.

[Sch66b] L. Schwartz, *Théorie des distributions*, Publications de l'Institut de Mathématique de l'Université de Strasbourg, No. IX-X. Nouvelle édition, entiérement corrigée, refondue et augmentée, Hermann, Paris, 1966.

[SD64] S. L. Sobolev and E. R. Dawson, *Partial differential equations of mathematical physics*, Translated from the third Russian edition by E. R. Dawson; English translation edited by T. A. A. Broadbent, Pergamon Press, Oxford, 1964.

[Str92] W. A. Strauss, *Partial differential equations. An introduction*, John Wiley & Sons Inc., New York, 1992.

[TS90] A. N. Tikhonov and A. A. Samarskiĭ, *Equations of mathematical physics*, Dover Publications Inc., New York, 1990, translated from Russian by A. R. M. Robson and P. Basu, Reprint of the 1963 translation.

[Vla84] V. S. Vladimirov, *Equations of mathematical physics*, "Mir", Moscow, 1984, translated from Russian by Eugene Yankovsky [E. Yankovskiĭ].

Index

Principles of Partial Differential Equations
DOI 10.1007/978-1-4419-1096-7

ERRATUM

Principles of Partial Differential Equations

Alexander Komech · Andrew Komech

© Springer Science + Business Media, LLC 2009

Erratum to: Principles of Partial Differential Equations
DOI 10.2007/978-1-4419-1096-7

The Digital Object Identifier (DOI) appearing on the copyright and chapter opening
pages of *Principles of Partial Differential Equations* is incorrect. The correct DOI is:
10.1007/978-1-4419-1096-7

The online version of the original article can be found under
DOI: 10.1007/978-1-4419-1096-7

Alexander Komech   Andrew Komech
Vienna University   Texas A&M University
Austria   USA